THE
PLUMBER'S
COMPANION

JAMES HASTINGS

THE

PLUMBER'S

COMPANION

DAVID & CHARLES: NEWTON ABBOT

0 7153 5461 2

Set in 10/11pt Baskerville
and printed in Great Britain
by Bristol Typesetting Co Ltd Bristol
for David & Charles (Publishers) Limited
South Devon House Newton Abbot Devon

CONTENTS

LIST OF ILLUSTRATIONS

Introduction

My plumbing apprenticeship began in Scotland with George Ferguson & Son of Airdrie, Lanarkshire, in September 1933. It wasn't that I had any particular ambition to be a plumber; jobs were hard to come by in those days especially in Scotland and any job would have done. But I took to plumbing as a duck takes to water; water was my element. And as water finds its own level, so any natural inaptitude was balanced by sheer enthusiasm. At evening classes many things began to make sense. The arithmetic, geometry, algebra, physics and chemistry which I could not grasp at secondary school clicked into place, as do the pieces of a jigsaw, and made a coherent shape.

I asked many silly questions and had my ears cuffed for asking too much. Questions like :

'Why is a bib cran called a *bib* cran?'

'Why is "ferrule" pronounced "virrol"?'

Questions like those were not answered.

It didn't occur to me to ask why a cran was called a cran probably because, like everyone else, I *knew* a cran was called that because it *was* a cran.

The ink has faded on my Scottish Certificate in Plumbing. But I can still make out the date—4 October 1938— and it demonstrates that I attended Coatbridge Technical College for five years. My apprenticeship was completed in September 1939 when I was twenty-one years of age.

9

The first thing that impressed me when I arrived in London in 1946 was the strange way in which the plumber there carried his tools; in a bag, of all things! My method of rolling tools up in a canvas tool-sheet and winding a tool-string round the bundle seemed quaint to the London plumber.

On the first job, the foreman said :

'Can you boss lead?'

'What do you mean?' said I.

'There are dormers to cover, can you do sheet lead work?'

'Of course I can work lead.'

I was expected to do the step flashings on the sides of chimneys in what seemed to me a tin-pot kind of way : cutting one strip of lead to a zig-zag shape and wedging it all into place. I preferred my own style of fitting separate steps like *real* steps, but I had to conform to the 'cockney' flashing.

There was difficulty with the language; I discovered that the English, in London anyway, do not speak very good English. When listening to someone I had to snatch from the gibberish a few words that were familiar and guess the rest. A plumber said something about 'plumber's union' and I, thinking he must be a shop steward, said : 'Yes' (or perhaps it was 'Aye') 'do you want to see my card?' He held up a boiler coupling and said : 'I mean this.'

Little did I know that the time would come when even I would think in such strange terms, and that I would forget many of the Scottish ones.

Being very much aware of how different the Scottish plumbing terms and methods were from the English I toyed, for some years, with the idea of putting both sides of the story on record within one book. Although I learned the trade the long way and was prejudiced against the 'trainee' system, I had to accept it, and my sympathy was with those young men who, having received a highly compressed plumbing course, went out to work in the many-tongued parts of Britain.

When I went as a raw Scot to London I had some confidence from a few years of experience, but even at that I was nervous; how would the trainee feel in similar circumstances? Perhaps if he had some kind of English-Scottish or Scottish-English dictionary as a 'companion' it might help.

Then, in the autumn of 1966, I came across a notice in the London *Observer* inviting applications for research grants from the Leverhulme Trust. With little hope of an ordinary, middle-aged plumber receiving serious consideration, I applied for a grant to help me with the field-work necessary for the compiling of an encyclopedia of plumbing, with particular reference to variations in local terms and methods.

As a working plumber with little academic background, I did not know that the subject of local variations in terms and methods was, in fact, original research, and that when one undertakes original research there is no knowing where it will lead. On the other hand, an encylopedia of plumbing can be compiled by using only second-hand, and even third-hand, material from previously published works. Some expertise in compiling could be more important to the compiler than first-hand technical knowledge.

My application to the Leverhulme Research Awards Committee was forgotten after a few days; my mind was taken up with a few sub-contract jobs for builders. The application had been made with a considerable degree of cynicism on my part, a cynicism engendered by many years of serving the British public which, in its ignorance, accepted as real the image of the plumber portrayed on television screens, in national newspapers and in cheap and unfunny music-hall sketches. Self-styled 'master builders' who took to themselves the grand sub-title 'sanitary engineers' could gull this public with fanciful chatter and, at the same time, live like parasites on the men who really knew—the plumbers. Architects can make pretty and convincing drawings but few include the plumbing. The architect says to his 'master builder': 'The plumber will be able to find his own way' and the 'master builder' nods. Slick salesmen have now joined the ranks of those 'master builders' in the parasite game. The potential customer is dazzled by the gloss of bathrooms, etc, with plastic veneer—no pipes—and is more concerned with 'How much?' than with a considered 'How?' While a few hundred pounds can be spent on expensive equipment for bathroom and kitchenette, the plumber has to struggle with pipes in impractical and nearly impossible positions to the accompaniment of constant bickering about the cost of his labour.

However, on the Saturday morning of the Easter weekend, 1967, I received a letter from the secretary of the Leverhulme Research Awards Committee which told me that I had been awarded £500 ($1,200) to help with my research.

To say that I leapt with joy would be telling a lie—reaction is always very much delayed with me. I was stunned and speechless for a few hours. Then there was an overwhelming sense of wonder that a mere plumber like myself should be so favourably considered. After that I was frighteningly aware of the enormity of the task I had undertaken; I did not know how to research, or even how to interview people, and I did not know how to handle the material once I had found it. Deep down there was the consolation that publication of a book was not a condition of the grant but, as I saw it, I had committed myself to deliver £500 worth of research and that meant publishing the results of my work. *The Plumber's Companion* is the result of research and personal experience.

The period of the grant was the year beginning 1 September 1967. Half of the grant was to cover loss of partial earnings and the other half an allowance for expenses when travelling. £500 is a lot of money, particularly to me, a plumber who had been doing for some time just sufficient plumbing to meet week by week commitments. I had no money of my own to finance research.

Una Long, my partner in a small one-man plumbing business in Fulham, London, instructed me on how to use index cards in collating my material instead of relying on scraps of paper and empty cigarette packets which, as most plumbers know from experience, get rather crumpled and indecipherable in the pockets of working clothes. Next she suggested that I should find out if there was any published work on the subject of local variations in plumbing; I spent a considerable time making inquiries at reference libraries. I could not find any previous work on my subject, but anyone who has been overawed by the vast Reading Room of the British Museum will appreciate that there could be such a book tucked away somewhere. How does a working man tread an academic field if there are no signposts? How can he get any depth if there is not some kind of sounding board? I could not have got anywhere with this work if Miss Long had not helped with what I call the mechanics of the job, and if she had not given me direction when I was lost. I am deeply indebted to her.

The fieldwork. I had to eke out the money over as much ground as possible—I had to travel cheap. Out on the road in Fulham was parked my 15cwt Thames van, a 1962 model, and known to me as 'the Cow'; her registration number is 80COW. I decided to live in her when travelling. First, she was reshod with four new tyres and she was given a good overhaul in a reputable garage.

Next, I fitted her inside with angle iron to support shelves on either side, and on the floor laid a carpet. Finally, the loading. On the top offside shelf were my changes of clothing, towels, and spare blanket. Below that were the foodstuffs, lots of tinned food included; a first-aid kit; cutlery; kettle; teapot; a gas camping stove. And below all that, the airbed, blankets, sleeping bag. Clean shirts, jeans, coats, etc, also hung on coathangers along that side.

On the top nearside shelf were my books of reference, including two volumes of the *Shorter Oxford English Dictionary,* Partridge's *Slang and Unconventional English,* Chambers's *Scots Dictionary,* and some light reading. Below that were my tape-recorder (a Philips) with spare cassettes; typewriter; a good stock of typing paper; camera; radio and a multitude of bits and pieces—just in case.

The bottom nearside contained a shovel, in case I became snowbound or mudbound; a stout rope, also in case; and a good, all-round kit of plumbing tools. The tools need explaining: I intended eking out the money by taking employment for a few days or even a week or two, here and there. Having done my homework, so to speak, as far as research was concerned, I was ready to take to the field by mid-March 1968.

It was 20 March when I set off from London to test my legs. My intention was to travel north on the A1, breaking off here and there on the way to seek out local plumbers—preferably elderly. Already I had learned that the plumbing experts had enough on their hands coping with standardisation and new methods to have much time for local terminology. Indeed, standardisation was designed to change the colour of local language to plain black and white.

Many times on my way north did I determine to break off at the next minor road, and as many times did my nerve fail—due to an inherent backwardness I have always been slow to speak with strangers. At last I made up my mind to stay with friends near Stockton-on-Tees, Co Durham, for a couple of days and to begin my inquiries in that area. On the first night there, I argued with myself that it would be better to start in Scotland amongst my own countrymen, but with a burst of spirit, next morning, I condemned any further procrastination and took the plunge.

Having made some timorous inquiries I soon found myself with an appointment to call at an elderly gentleman's home in Stockton-on-Tees—he was a retired plumber. With notebook in pocket and tape-recorder in hand I presented myself, dead on time, at the front door, apprehensive about the forthcoming interview.

He met me with a shake of the hand. We went through to the sitting room where his wife said: 'Come in, love, and make yourself at home.' They little knew what that welcome meant to me. It broke down the first of the terrible barriers I had been faced with. The wife disappeared and left us to talk—and we did talk. For two and a half hours we discussed plumbing—critical of modern methods, of course—and occasionally I thought that if I was anything of an interviewer I should be guiding this conversation along properly defined lines. But I didn't; we were simply talking plumber to plumber; sheet lead work on roofs, on domes and turrets; lead pipes and solder joints; copper pipes and iron pipes; soil pipes, waste pipes and drains; rainwater heads and gutters; wells and pumps; pewter and tin, tin beer-pipes and beer machines. My tape-recorder used up two sides of the tape—he didn't mind speaking in the company of the tape-recorder. Although not much of an interviewer, I felt that I had broken the ice. There were times when we had to make sure we were speaking about the same thing, as when I brought up the subject of the pot, meaning solder pot. He looked at me rather strangely: 'You mean one with a P-trap or an S?' And when I said the one for melting solder he said: 'Oh, you mean the metal-pan.'

From Stockton-on-Tees I went on to Scotland and for about two weeks I stayed with friends in Lanarkshire, travelling out each day in all directions in my quest. I still had difficulty in approaching people and if they were too busy earning a living, as they often were, to spend any time talking, I was easily discouraged. A sense of urgency got into me and I felt that I was not getting anywhere very fast.

Then, having overstayed my welcome I am sure, I set off through Ayrshire to Stranraer, Dumfries, and across southern Scotland. The hours of daylight having lengthened, I had got into the habit of bedding down with nightfall and rising shortly after daybreak; the nights were cold but the weather fine.

Although there were full tapes and considerable notes to show for this time, when I arrived back in London in mid-April I was unhappy with my progress. A week was spent in London, taking stock of the situation, doing some practical plumbing, and getting the card index up to date. On 22 April I dashed off south to Kent—dashing here and there all the way.

Now it happened that, weary in mind and in body, I sat on the tail of the van one evening near Old Romney on the south coast of England. Parked for the night, I was just having a bite to eat and a cup of tea. Perhaps it was the stillness of the evening that accentuated the exhaustion, but suddenly, out loud, I said:

'Where in Heaven's name *am* I going? Where?'

And equally spontaneously, as if the words had been put into my mouth:

'Nowhere. You're there.'

A great load seemed to lift from me as I relaxed, body and soul, in a wonderful awareness of the halted moment.

With newly-awakened senses, I really saw the red, falling sun and the rising mist from the marsh, and there were sheep with bleating lambs, and birds were having a final song before they, too, bedded down. I thought to myself that if nothing more than this awareness came of my research, my journey would be well worth while; the Leverhulme Research Awards Committee would have given me more than they ever knew. I went for a long walk before darkness fell.

In this new-found frame of mind I had a quiet look at the question of research and my attitude to it. My false modesty, which I had always called shyness, was the greatest problem; I must remember the falseness of it and try to replace it with some feeling rather of humility when approaching people; strangers were just people.

Una Long had told me how easy it would be to find only what I wanted to find, and it would be easy to bend facts to what I wanted to find. Now I saw what she meant. Objective research would mean that a negative finding would have some meaning; until now I had been disappointed when I found nothing new, and felt that I had failed in my mission.

I also realised that my nature would not allow me to work with the tools in random places for random periods. It would be like working on two different levels at the one time, and if I tried to do that, one or both would suffer. Anyway, with the open-road feeling I couldn't settle to a job. I decided to do my fieldwork for a few weeks at a time and between those times go back to London to earn a living and, if possible, do a little bit more to further my research.

It was in that frame of mind that I settled down to the remainder of my fieldwork. It was not easy to change my ways, but all the way across the south to South Cornwall and back to London I managed to maintain the right approach for most of the time. There were lapses of panic

B

and uncertainty but they did not wholly control me as before.

People have said : 'Weren't you afraid when living alone in the van? How did you sleep at nights?'

Of course I was frightened at times. One night near Salisbury there were stealthy footsteps near the van, then a violent banging on the sides. I drew back my curtain and was blinded by a torch light. It was the police; as I was parked near a deserted farm, some local person, thinking the gipsies had moved in, had phoned the police. Another time, in a gravel magazine near Appledore, Kent, in the early hours of the morning, a bunch of young fellows drove in beside my van; there was a hilarious time of shouting, swearing, beer drinking and sporting about the gravel, but they did not touch the old Cow. I always kept my dummy or dolly (see p 67) near at hand at night.

There were also false alarms. Like the time in Northumberland when I was awakened by a bumping and scraping and rocking of the van; a bunch of cattle were having a lovely time rubbing themselves affectionately against *my* old Cow.

Towards the end of 1968 I gave the secretary of the Leverhulme Research Awards Committee a summary of my work during the period of the grant. The Committee was, to my surprise, quite impressed. As there was still much to be done, I applied for an extension of the grant and in due course, on 31 March 1969, I learned that I had been awarded a further £500 ($1,200) on the same terms as the original grant. That further help allowed me to do some more travelling in the northern parts of England and in Wales and the south of Scotland.

In 1968 I was greatly relieved to be able to move away from the din of London to Ely, Cambridgeshire, where I now live. That meant I could travel to any of the regions still to be explored without getting involved in London traffic. I have been able to go out for single days at a time, travel considerable distances, and make extensive inquiries. I found that my travelling, when away from home, was best done during the early or the late summer; many of the camping sites were convenient and empty and thus cheaper, the owners often charging merely nominal fees.

When not travelling, much of my time has been taken up by corresponding with kind people who contacted me because of publicity I received in the national press, on radio, and on television. The plumbers who wrote to me were most helpful as were those who kept up correspondence after contact during the fieldwork.

I started writing *The Plumber's Companion* on 14 December 1969. By the end of April 1970 it was finished except for a few thousand words which I intended to add when my 1970 fieldwork came to an end. When I returned to Ely in August I did not like the shape of the book and put it out of sight while I got on with some manual work.

Meanwhile the publisher asked if there were any terms in the United States of America which would be of interest. I have never been to the States and doubt very much that I ever shall go, but I made a few inquiries and that led to further correspondence with contacts in America. The result of that correspondence has been kept separate in its own Supplement at the end of the book, mainly because the book is primarily about the local variations in Britain. I trust that the United States glossary will be added to as time goes on; there is still a great deal to be done in this field.

Many of the terms I came across made me look to previous publications. The first to fall into my hands was W. P. Buchan's *Plumbing* (6th edn) which I was able to borrow from a gentleman in Cardiff until my public library obtained for me a seventh edition.

The public library was able to get for me the three volumes of P. J. Davies' *Standard Practical Plumbing* and also S. Stevens Hellyer's *Lectures on the Science and Art of Sanitary Plumbing* and *The Plumber and Sanitary Houses*.

Often have I heard plumbers in London mention 'touch' and any time I have seen the word in print it has always been in inverted commas. But the only explanation I could ever get was that it is used to touch the part to be soldered. On that pretext, my logic tells me, any tool could be called touch. Davies said : 'Touch is a tallow candle'; it is not, and this is a good example of the help I received from old books.

Davies' reference to 'tallow candle' inspired me to make some inquiries from Price's Patent Candle Company and other firms dealing in animal fats. Price's, to my surprise, stated that they still made plumbers' candles and kindly supplied me with one of each type, although they had never heard the word 'touch'. However, my *Shorter Oxford English Dictionary* and my Chambers *Scots Dictionary* gave me the answers regarding the meaning of 'touch' (see pp 151-2). This research programme has taught me how to make everyday use of my dictionaries to glean information.

Having made extensive use of books, coped with letters to and from contacts in Britain and in America, and got rid of some weight by dint of hard physical work, I settled down to rearrange and rewrite *The Plumber's Companion* on 19 October 1970. As far as the drawings are concerned, in the main they are my own work, and I make no apology for their quality. I have not done any drawing since night school days back in the thirties.

Those who would standardise everything may hold up

this book of variations in terms as an example of the confusion caused by lack of standardisation; they would eliminate dialect. Men from the West Country and Wales have suggested that we should forget these 'slang' terms, and in this age when plumbing is a precise and technical science we should have everything—and everbody— standardised. If a Welsh man is willing to eliminate his dialect, how soon before we have the Welsh sacrificing their language on the altar of standardisation; standardisation knows little restraint.

In the introduction to his book *Lectures on the Science and Art of Sanitary Plumbing,* S. Stevens Hellyer wrote in 1882 :

Men are so different, you cannot level them like water. If all men were cast in the same mould, like rain-water heads of a building, they might have the same treatment; but as they are as various as the chimney-pots of London, they must have various treatment . . . . It is just the same in the natural world. Come with me to the cliffs of the sea, and watch the wild waves marching to the shore. How they differ in size, in height, in sweep! Linger here awhile, until the fall of night, and then lift your eyes, and behold the night-lights of the sky, and see how they differ; and as one star differeth from another star in glory, so does man from man here and everywhere. . . .

The chimneypots of London are disappearing and the horizon is becoming a featureless, concrete line. Men are being trained to think in terms of British Standard numbers, and, with no original thought, are marching off the production line like robots or like plastic rainwater heads all made to one standard.

In recent years the skill has gone out of plumbing; many plumbers know nothing of sheet lead work and could not make a solder joint on a pipe—they do not have to. I have felt that I am old-fashioned, that I should be keeping pace with the plastic trend. I could not try to keep pace because I have felt also that there is no place for me in the rat-race. I have wrestled with myself to find out what was wrong with me. Was I holding to the old for the sake of the holding, or for the sake of the old? In industry I saw the grasping and the striking, in people in general the discontent and, all around, the punch-ups and brashness; I felt that I didn't belong. 'Why?' I asked myself, and then : 'Is this the first sign of old age?'

One day when I was making a lead fountain-head for the garden the answer came to me. I remembered vying with other apprentices to see who could make the best solder joint or who could show most skill in shaping lead. There was deep satisfaction in making something well, something that would last. When a nice job had been done, we used to exclaim : 'A thing of beauty is a joy for ever' and though we joked, we meant it. There was fulfilment. Now I look at my lead fountain-head—a six-rayed sun face—watch the water piddling from the mouth, and know that nobody will ever quite copy that head; made with my own hands, it has a character all its own and not even I could make an exact copy—not quite the same. There is fulfilment.

In writing this book I soon found that I had to make a set of rules to guide me when flummoxed. On the wall at the back of my desk I can still see them : 'Name it' (main heading); 'what is it?'; 'what's it for?'; 'origin of word?' Those rules for the outline of each definition have been bent, stretched, broken and sometimes ignored. For instance, I have started this book of variations in plumbing terms with 'AIR LOCK' for which I have not found an alternative word or phrase but it is included for the benefit of any layman who cares to read. In the past, many housewives who complained of 'air knocking in the pipes' have been upset when, instead of 'clearing the pipes', I have put a new washer on a tap or re-packed a stuffing box to get rid of water hammer.

For instance, again, with 'WATER CLOSET' and 'WATER WASTE PREVENTER' I have gone on to present short histories of their evolution.

The information in the main text has been gleaned from England, Scotland and Wales (those are given in alphabetical order and not in order of preference); I regret that the troubles in Northern Ireland made it impracticable to do research there. In looking for a particular word or phrase, the reader should refer to the Index.

I present *The Plumber's Companion* knowing that it is not a complete vocabulary for plumbers—there were some two thousand terms in my file before I weeded out those with no local variants—but I hope that this small book will serve to stimulate interest in words which, if not collected and put on record, will be lost for ever. I should be pleased to hear from anyone who comes across any inaccuracy in my work or who knows of any local words or phrases which have been omitted but which should have a place in a book of this kind. Indeed information would be welcome from other English-speaking countries regarding local plumbing terms.

James Hastings
November 1971

The Plumber's Companion

Note: Names of authors used in the text indicate that the information has been taken from a book included in the List of Sources—eg 'Buchan' refers to W. P. Buchan's book called *Plumbing*.

Other abbreviations are: BS = British Standards; OED = Oxford English Dictionary; MAC = Metal Agencies Company; all of which are fully detailed in the List of Sources. Cross references to the Supplement of terms used in the United States of America are indicated by 'US': eg 'See also US BUFFER PIPE'.

AIR LOCK

Such strange ideas does the layman (and laywoman) have on the subject of air lock that, although the term does not have any local variants, it is worthy of mention if only to point out the difference between it and 'water hammer' (qv) with which it is so much confused.

When a tap coughs and splutters or even stops running, it is likely that the cause is air in the pipes. This is known as air lock. It can come about in supplies from storage tanks, hot or cold, but if the hot tap has an incurable cough it may be that the hot supply branches off the expansion pipe at too high a level, and such a state of affairs can be cured only by some alterations in the pipe runs.

A novel way of freeing air-bound pipes was once used in Glasgow tenements. The tenements were supplied with water from huge lead-lined tanks in the roof spaces and lead pipes snaked out from the bottoms of the tanks, sagging their up-and-down way over the ceiling joists. As the pipes were crystalline with age the plumber dare not disturb them in any attempt to straighten out the sags. Obviously air would collect in the hummocks here and there and thus air locks were common. When called upon to clear an air lock, the plumber would pierce the offending pipe with hammer and nail to release the air; he might make several holes to find the right hummock. Having cleared the pipe of air, he would tap in a small tapered wooden plug to seal each hole. In a pipe which had had such treatment, any further air locks could be cleared by taking out a plug or two and then tapping them back into the holes. A plumber who was on regular maintenance work knew his area and was familiar with the locations of plugs in various properties. Such a practice is not recommended.

AIR VESSEL

BS 4118 gives: 'A closed chamber which utilizes the compressibility of contained air, either (i) to promote a more uniform flow of water when connected to the delivery pipe or suction pipe of a reciprocating pump, or to the delivery pipe of a hydraulic ram, or (ii) to minimize shock due to water hammer when connected to a high pressure water system.'

'Air vessel' and 'air chamber' are the usual terms in England and Wales. Water hammer is a common domestic complaint in those places, probably because plumbers do not seem to be air-chamber-conscious; the most common method of connecting a high pressure supply to a ballvalve, for example, is horizontally with a straight solder joint or with a straight or bent copper coupling.

Air Cushion

This Scottish form of the term 'air vessel' suggests that the Scots are well aware of its purpose: to act as a cushion to absorb shock in pipes. When jointing a main supply to a ballvalve, a Scottish plumber will almost automatically do so with a branch joint so that a section of the supply stands vertically higher than the branch; the top of the upstanding pipe is sealed and thus air is trapped to form the necessary cushion. (See also US BUFFER PIPE)

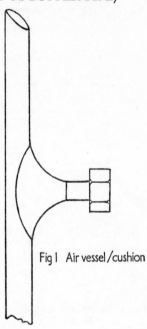

Fig 1 Air vessel/cushion

APRON

BS 2717 defines an apron flashing as 'a flashing the lower edge of which is lapped over the roof covering'.

That describes the English and Welsh apron but not the Scottish. The OED describes the flashing which the Scots know as an apron as a strip of lead which conducts the drip off the wall into the gutter. The compilers of the OED could have said that the strip may be of an impervious material, usually metal, and not only lead. Also, the Scottish apron, known in England and Wales as a 'string flashing', is used at the front of chimney stacks to conduct the drip off the breast on to the berge, which is known in the southern counties as an apron, on to which it laps. In English terms the sketch shows the section of the front of a chimney stack with string flashing lapping the vertical part of an apron or, in Scottish terms, the apron lapping the vertical part of a berge. (See also BERGE)

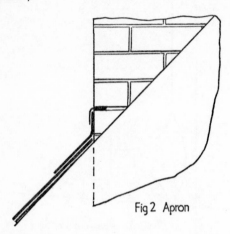

Fig 2 Apron

Stepped Apron

That the Scots see an apron as something which laps the vertical part of another flashing is emphasised by this term used by Buchan for what is known today as a step flashing. The Scottish single step flashing always laps the vertical part of another flashing—like the upstand of the side gutter of a chimney—and thus it is indeed a stepped apron. (See also STEP FLASHING)

ARRIS GUTTER

Known nowadays as a gutter which one might fit to the eaves of a garden shed or garage and which, being made by fixing lengths of wood to form a V-shape, is not plumbers' work. But very old buildings and ancient monuments often have arris gutters on a larger scale than the garden shed type and which were lined with lead, zinc, or copper—so the plumber had his part to play in such work.

In Scotland such a gutter would be a rone (qv).

BACKNUT

The locknut on the screwed stem of a tap or pipe fitting for securing it to some fixture—eg for securing pillar taps to baths and basins, the backnut screwing up the stem on the underside of the fixture.

Also, a backnut, dished on one face to retain a grommet, is used in the making of watertight joints to tanks, and in conjunction with long threads and connectors.

The word is known throughout England and Wales, and in parts of Scotland, but in Cambridgeshire the backnut was also referred to as a locknut. (See also JAMNUT)

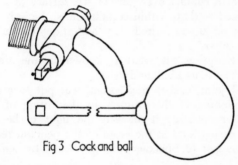

Fig 3 Cock and ball

BALLCOCK

Although now wrongly used, as are ballcran and balltap, with the meaning ballvalve, the original ballvalves really were cocks (see VALVE). The cock was a horizontal bib cock with a square head, the square head being filed down to fit the square hole on the end of the arm attached to the ball. As there was a danger of the device locking when the tank or cistern was nearly empty, the cock was knighted; that is, the shell of the cock was grooved and a pin inserted in the spindle or plug so that the arm could not fall any lower than the knighting allowed.

The Scottish ballcran could well be used to describe that type of ballcock because, having a bent nozzle for an outlet, the cock was indeed a cran.

BALL HAIR

Slang term with reference to a vague and very slight measurement. As the ball part of the term is meant to emphasise the lightness of the thickness in question, one is left to assume that it is finer than, say, the hair of the head. (See also KNURL)

BALL JOINT

BS 4118 gives : 'A joint in which the two parts are shaped so as to be in contact over part of the surface of a sphere and arranged so that, within certain limits, the axis of each part can be set in any desired plane and at any desired angle, one to the other.'

That description fits what is known in the trade, in Scotland at least, as a ball and socket.

The ball joint, or ball and socket, was put to great use at the beginning of this century in the fitting of ceiling pendants for gaslight; it allowed the light to be moved slightly as desired and in the event of the pendant receiving a knock, danger of breakage was prevented by the movement.

Solder Joint

The underhand solder joint is known in Scotland as a ball joint.

BALLOON GUARD

May be a *wire* balloon or *wire* ball (qv), but since the coming of plastics, the balloon or ball may be of plastic and the 'wire' no longer applicable; 'balloon guard' covers any material.

BALLVALVE

BS 4118 describes a ballvalve as: 'A valve, for controlling the flow of water into a cistern, which is operated by a lever arm with a float attached, the float riding on the surface of the water in the cistern.'

Then follow descriptions of specific patterns: Croydon type, delayed action, diaphragm, equilibrium, Portsmouth type, and reverse action.

The Croydon, controlled by a vertical piston, and the Portsmouth, controlled by a horizontal piston, are the types most used in domestic plumbing; the Croydon is better known in the south of England than in the north. In recent years there have been many variations in design, and plastics now play a great part in the manufacture but the principle '. . . operated by a lever arm with a float attached . . .' is the same as before.

If a householder has reason to complain that a closet cistern makes too much noise when filling or that it takes an uncommonly long time to fill, the fitter of the cistern has failed to consider the water pressure in the supply. Ballvalves are described as 'high pressure' (HP), 'medium pressure' (MP), 'low pressure' (LP), and 'full-way' (FW). A high pressure ballvalve under low pressure will provide a mere trickle of water. A full-way ballvalve under high pressure will give such a rush of water that the cistern may develop continuous flush because the emptying flush is less than the incoming water.

Armitage Ware Ltd have provided, for a number of years, a pamphlet which gives a table of suitable pressures for types of ballvalves:

HP	over 45lb/sq in or over 100ft head
MP	20–45lb/sq in or 45–100ft „
LP	10–20lb/sq in or 20–45ft „
FW	under 9lb/sq in or under 20ft „

In the case of a high pressure ballvalve under high pressure, the noise problem may be got over by fitting a control stopvalve on the supply to the cistern.

Jennings's patent, in 1852, was probably one of the earliest ballvalves of what might be called the diaphragm type. (See WATER WASTE PREVENTER)

Davies describes a ballvalve as 'a faucet governed by a valve, a kind of ball cock'. His use of the word 'faucet' (qv) in this sense indicates that it was still used in the old English sense, now adopted by the USA.

Ballcock (see BALLCOCK)

A misnomer used in many parts of Britain for the ballvalve of today. The ballvalve is not a cock.

Balltap (see also BALLCOCK)

Quite a common word for the ballvalve, used mainly by the layman who probably feels that 'tap' is not quite so vulgar as 'cock'. The ballvalve is not a tap.

Ballcran (see also BALLCOCK)

A Scottish misnomer for ballvalve—it is not a cran.

BARREL

Nowadays 'barrel' is mild steel tube and, though the term was used more widely during the last hundred years, it is now confined to the London area. In the days when plumbers used wrought iron pipe that, too, was barrel. There is black barrel, the heavy quality for hot water systems, the light quality for gas known as gas barrel, and there is galvanised barrel for water supplies and, very often in certain areas, for waste pipes. The heavy barrel once painted red to denote 'steam' was 'red steam' or 'steam barrel'.

Britannia Iron and Steel Works at Bedford (the well-known GF—George Fischer—fittings are made there) provided part of the story of 'barrel'.

William Murdock (1754–1839) a Scottish inventor born near Auchinleck, Ayrshire, discovered how to extract gas from coal for the purpose of illumination and as early as 1792 was able to light his own cottage and office with gas; that was in Redruth, Cornwall. In 1802 a portion of his

Soho factory in Birmingham was illuminated by gas to cele-
brate the Peace of Amiens.

Although gas was then available, the great problem con-
fronting the early gas engineers was how to get it to where
it was wanted; steel and wrought iron, involving manufacturing processes of great complexity, were expensive.
Murdock got over his immediate problem by buying vast
quantities of what we would now call 'war surplus'; he
bought musket barrels left over from the Napoleonic wars
and welded them end to end to make convenient lengths.
Thus 'barrel', but the term is now extended to mean pipe for
water and steam as well as for gas. In recent years, 'copper
barrel' has been heard in London with reference to copper
tube, but in this context the original meaning is lost.

Lead Barrel

There is no connection between Murdock's 'barrel' and lead
barrel once used in the making of lead jack pumps. The
barrel of a jack pump was the section which housed the
valves and the vertical rod attached to the handle; in the
author's Ely museum there is a cast-lead pump, dated 1799,
with barrel 4in in diameter and 4ft high. Grey & Marten's
1923 catalogue lists lead barrel of diameters varying from
$2\frac{1}{4}$ to 12in and also lead soil pipe of diameters from 2in to
6in. The 4in lead barrel, for example, could weigh from
48lb to 108lb per yard while 4in lead soil pipe would weigh
a mere 17lb to 42lb per yard.

Nowadays a plumber would be unlikely to come across
lead barrel unless he was asked to repair an old lead jack
pump.

BARREL NIPPLE

A very short piece of barrel with a male thread at each
end. Because of the inconvenience of threading such small
pieces of pipe, plumbers usually find it more economical to
purchase barrel nipples from the merchants.

C

Running Nipple

Although not usually referred to as a barrel nipple, the running nipple is indeed just that, as it is made by threading a piece of barrel to a considerable length and cutting off the length of threaded pipe as required.

BAT

A lead wedge for securing lead flashings to masonry or to brickwork. The flashing edge is turned into a chase or raglet and the bat, driven into the raglet, wedges the flashing into position. 'Bat' is the usual name in Scotland and the northern parts of England but in other regions of Britain plumbers are not familiar with 'bat'; they know it simply as 'lead wedge'.

The word goes back to medieval times meaning a piece or a lump, and in that sense is perpetuated by bricklayers who still speak of pieces of brick as bats; the size is usually declared as 'a quarter bat' for a quarter of a brick, 'a half bat' etc but never a whole bat.

But 'bat' also means a blow—a bat-on-the-lug—so it is difficult to determine if the lead wedge is a bat because it is a piece of lead, or because it is driven home with heavy blows. The driving in is called batting and is done with a batting iron (qv) and hammer, *not* with a bat. (See also THUMB BAT)

BATTING IRON

A tool similar to a cold chisel except that what would be the cutting edge is blunted to a thickness of up to $\frac{1}{4}$in so that when driving in bats (qv), or batting, it really drives and does not merely cut into the lead bat. As 'bat' is a Scottish word in plumbing, so also is batting iron.

In South Wales this tool has been spoken of as a drift (qv). (See also CAULKING TOOL)

BAY WINDOW

A projection of a house, forming a bay in a room and filled with a window arrangement. The plumber may be called upon to cover the roof of the bay with sheet lead, zinc, or copper, or to fit flashings to weather a slated or tiled bay. If the bay window is on an upper floor only, supported from the ground or by corbelling, it is an oriel (qv).

BEADS

In the bending of lead soil-pipes and waste pipes, short cylindrical pieces of wood, usually box, are frequently pulled or pushed through the pipe as bending progresses; these pieces of wood are known as bobbins and followers. Beads of rectangular cross-section were used in bending rectangular lead rainwater pipes. One method of using them is to string them on a length of sashcord and pull through; thus the term 'beads' used in places as far apart as the south of England and Northumberland. The leading bead, the bobbin, has a diameter suitable for the bore of pipe being bent and the 'followers' slightly smaller.

An old method of driving the bobbin was by using numerous 'followers', without cord, and driving them from one end with a mandrel (qv). In Scotland, 'followers' are little known, the method there being to use one bobbin on a cord and with a heavy hammer with a lead head, tug the bobbin from one end to the other. Very little bending of lead pipes is now done, for we live in the plastic age. The beads were superseded by patent bending springs, probably in the early 1920s, but most plumbers, particularly in England, stuck to the beads.

Very old plumbers speak fondly of the old days of the ancient art of pipe bending with beads, and shake their heads with despair at the new-fangled bending springs and the use of plastics. But it was only in 1882 or thereabouts that Hellyer, Davies and Clarke went through the same motions of despair regarding the evils of the new-fangled bobbins or balls, insisting that the correct method of pipe

bending was with dummy and/or bending irons (see BENT
BOLT). Also in 1882 Hellyer said: '. . . bending of lead
soil and funnel pipes (without cutting or soldering them)
commenced about forty years ago' which means that pipe
bending goes back no further than, say, 1840—not such an
ancient craft. Before that time, soldered elbow joints were
used when a change of direction was necessary in a length
of lead pipe.

BELL HANGER

In the north of England and in Scotland one often sees a
trade sign 'Plumber and Electrician'. Such a seemingly
strange combination of trades is easy to follow in the evolution of the trades.

Before electricity, plumbers used to install elaborate
systems consisting of bells, cranks, copper wires and pulls
throughout houses; in old houses there are systems of wires
still in existence. The plumber had to fix the cranks and
wires so that there would be free movement and he had to
make small sleeves, usually of zinc, to conduct the wires
through walls, etc. The bell-board was normally in the
kitchen quarters where the maid, or butler, could see the
tell-tale flap which signified which bell had rung.

When, in 1868, Georges Leclanche devised a method of
generating electricity in glass jars—known to many old
plumbers as Leclanche jars or Clanche jars—the job of
fitting up electrical bell wiring systems remained with the
plumber. The jars had to be refilled with distilled water
periodically, but if reminiscences can be believed, many
plumbers did not trouble to fetch *aqua pura*—they made
their own water, especially if the jars happened to be situated
in the roofspace out of sight.

Most plumbers relinquished their advantage in the electrical field in the early stages; electrical engineering became a
trade in itself and plumbers were left out in the cold;
plumbing covered such a large field, which was becoming
more technical, and coping with electrical work was probably too much for small firms. The progressive firm, in time,

became two in one: a plumbing department, and an electrical.

BENT BOLT

Davies described what he called the bolt, the pin, or Tommy, etc—but he did not give any detail of the 'etceteras':

A bent piece of steel pointed so as to enter and enlarge a bored hole in a pipe for fixing cocks or bosses. Also used for pipe bending.

Many plumbers will recognise that description of what they call a bent bolt. In the southern half of England and in Wales, bent bolt and bent pin are the usual terms. There is little doubt that those terms have arisen because the bolt is indeed bent—it has two bends. But proceeding north the term changes, and 'bent' becomes 'bending'—for instance, in Scotland 'bending pin' is the standard form, sometimes 'bending iron'—but always 'bending'.

When Davies remarks that it is also used for pipe bending, he does not mean the kind of bending for which it would be used today, like sticking the bending pin into the end of a lead water pipe for leverage to bend the pipe. In his day, before bobbins (see BEADS), the bolt was used in the bending of small bore waste pipes, from $1\frac{1}{4}$in to 2in, in much the same way as the dummy has been used in recent years for bending lead soil pipes.

With the bolt inside the pipe, the heel of one bend on the dent to be knocked out, the bolt was hit sharply with a hammer.

Bent bolts of today measure from nine to fifteen inches from tip to tip, but when the bolt was a tool for bending in the old style, the length would be about thirty inches for the long bolt and nine inches for the short one.

Thus, while the term 'bent bolt' describes the shape of the tool, 'bending bolt' can be a reminder of historical interest in pipe bending.

BERGE

A Scottish term, variation of barge, meaning the main flashing on the breast of a chimney. The berge is called an apron in England and Wales. (See also APRON)

Buchan uses 'barge' with reference to the front flashing of a chimney which has one side flush with a skewed (see SKEW) gable. The OED gives barge-stones as stones forming the sloping line of a gable. Over the years since Buchan's time, it would appear that 'berge' or 'barge' has come to mean the main flashing on any chimney breast; a chimney stack which is positioned astride a ridge, so to speak, would have two breasts and thus two berges.

BIB TAP

The layman will know this as the tap over the kitchen sink. BS 4118 gives : 'A tap with a horizontal inlet and a nozzle bent to discharge in a downward direction.'

This BS also says that the term 'bib cock' is deprecated, but more often than not 'bib cock' is used with reference to bib tap in plumbing conversation. As a tap (in the true sense) is a cock, there seems no reason to deprecate the cock part of the term. Plumbers frequently abbreviate the term to 'bib' but although the question : 'Why is a bib called a bib?' has been put to many plumbers throughout Britain, an answer has not been given. The reaction has been 'What a damn silly question!' (See also US BIBB)

The answer may lie in the BS definition : '. . . discharge in a downward direction . . .'; perhaps not; a bibtap does not have a bib. The OED gives 'bib' as perhaps from Latin *biberer*, to drink, to tipple; that was in the Middle Ages. The conclusion is that a bib tap is a tap from which one would bib, or drink, and that would fit the bill in the days before hot water was taken to taps and before cold taps were taken off storage tanks (as in the London area); one should not drink cold water from tanks or hot water from taps. But nowadays, sensical or nonsensical, there are hot bibs as well as cold bibs. In certain circumstances, where there are three taps over a sink or an extra tap beside a basin, one of them is marked DRINKING to indicate that the supply comes direct from the main. How much easier it would be to print BIB on a small porcelain disc on top of a bib tap.

BIDET

Partridge gives Grose's definition of bidet or biddy : 'A kind of tub, contrived for ladies to wash themselves, for which purpose they bestride it like a little French pony or post horse, called in French bidets.'

The earliest record of 'bidet' in the OED is 1630.

Grose handled very delicately the reference to the bidet's purpose, but BS 4118 is more down to earth : 'A pedestal sanitary appliance, on which the user sits, for washing the excretory organs.' The user can be of either sex, of course; the use of a bidet can be refreshing and soothing to people, male or female, with certain complaints in the lower regions. It is also an ideal appliance in which to wash the feet.

However, it would appear that a few plumbers have a notion that the bidet's prime purpose is the washing of the feet. When a lady in London area complained that the bidet had been installed too near a side wall, the plumber retorted : 'I don't know what you're on about, there's plenty of room to wash your feet.' Later, the lady said : 'It was not *my* place to tell him what it is for.'

Because of its resemblance to a closet pan, the bidet is too often illustrated in catalogues alongside the pan, and one is reminded of the newly-rich who bought two pianos for the parlour because they were such a nice match. Perhaps because of this close relationship the idea that the bidet is a quaint French curiosity is perpetuated.

Working-class (for want of a better phrase) spelling and pronounciation was 'biddy' not so many years ago, particularly in the north, but the French with-it influence, having pervaded London and the south east, has crept northward, and French spelling and pronounciation is general. A few old gentlemen have been heard to pronounce it 'bidette'.

BIRD'S EYE

When bossing sheet lead the skill of the plumber lies in keeping the lead of a uniform thickness throughout and if it seems to be thinning he must dress the lead in such a fashion as to thicken where necessary. Should he allow the lead to become thin and subsequently split, that split, because of its appearance, is called by many old plumbers 'a bird's eye' or a 'cock's eye'.

BIRD MOUTH

The ornamental finish to the end of an overflow pipe, cut in the shape of a bird's open beak. Bird mouth and bird beak are fairly well known in England and Wales though the bird-mouthing of overflow pipes is seldom practised nowadays.

When a lead pipe has to be soldered to the corner of a cesspool, say, the pipe is first cut to the bird mouth shape to fit the corner before soldering. Any pipe cut to this shape is a bird mouth or bird beak.

BLADDER

An inflatable bag, used in pipe testing, which, having been placed in the pipe, is inflated to seal that end of the pipe—bag stopper is the standard term.

A plumber in East Anglia, speaking of old times, told how plumbers found a pig's bladder more effective than the manufactured canvas bag. Having collected pigs' bladders from the local slaughter house, the bladders were cleaned and pickled in salt to preserve them. The bladder could be inflated with a hand pump and a rubber tube when stopping a pipe.

BLEEDING COCK

When confronted with a main stopcock of the plug type, an Ely employee of the local Water Board immediately pronounced it a 'bleeding cock or a bleed cock'. The descriptive word was not an expletive although the gentleman did not know why the item should be a 'bleeding cock'.

It is possible that some plumber in the area became familiar with the Scottish frost cock (qv) which he might have been told was for 'bleeding' dead water from a pipe. Because of the similarity of the frost cock with an ordinary plug cock, every plug cock could have become, in his mind, a bleeding cock. By passing from mouth to mouth the term would soon become common local usage.

BLISTER

Heard in Lanarkshire, Scotland, for a solder patch on a lead pipe; a solder patch looks like a blister on a pipe.

When a lead water pipe bursts due to frost, a section of the pipe, about twelve inches on each side of the burst, should be cut out and replaced with new; it is said that weakening extends for at least that distance from the point of actual bursting.

The blister should be used as a temporary measure only, until such time as renewal can be carried out. Plumbing has deteriorated so much, particularly in England and Wales, that the patch has become recognised as a permanent repair. But should that same section of pipe freeze up again, it is almost certain to split alongside the previous patch.

BLOWLAMP

When a plumber refers to his 'lamp' he means his blowlamp. Modern blowlamps are wholly based on bottled gas, but in the seventh edition of his book Buchan mentions it as a gasolene blower.

In 1882 the first adjustable blowlamp was invented by C. R. Nyberg, a young Swedish technical genius who preferred design work to business and subsequently entrusted the selling and manufacture of blowlamps to the well-known machine experts, Max Sievert.

Fig 4   Type 'S' of 1882 (prototype
for all blowlamps)

British plumbers were slow to take to the blowlamp and
many stuck to the old methods of wiping solder joints well
into the twentieth century.

Fig 5
Self acting blowing lamp

Self acting Blowing-lamp

In 1882 Hellyer reckoned that men employed by his firm
had been using this lamp for many years (he does not say
how many). It was a French invention.

A boiler in the top section was half filled with methylated spirits and a small tube from the boiler led to a jet half way down the case of the lamp. A small spirit lamp was placed in the case, the burning wick immediately under the boiler. The gas generated in the boiler was forced by its own pressure out of the jet and, passing through the flame of the spirit lamp, provided a fairly long flame. If the safety valve on top of the boiler was not kept in good working condition there was some danger of an explosion. According to Hellyer, very nice wiped joints could be made with this lamp.

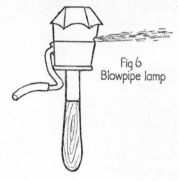

Fig 6
Blowpipe lamp

Blowpipe Lamp

The self-acting blowing lamp, perhaps because of its quality of danger, was not widely used, most plumbers with an ambition in that direction preferring to do what they could with the blowpipe lamp. This type of lamp was simply a tubular container with a thick wick for burning methylated spirits. Attached to the body was a brass tube to which was fixed a rubber tube; the narrow end of the brass tube could swivel and be directed at the burning wick whilst the operator blew air by means of the rubber tube. A fine long flame was the result. The earlier type of blowpipe lamp did not have a wick, merely a cup to contain burning spirits. Hellyer writes of a larger size of blowpipe lamp with which a solder joint could be wiped with the solderer or mate doing the blowing if a pair of bellows was not available.

The small blowpipe lamp was and is still used for making

'blown' solder joints. Before such a lamp, a simple blowpipe was part of the plumber's kit.

Blowpipe

The blowpipe of the nineteenth century was a trumpet-shaped copper tube, about nine inches long, with the thin end bent round, and an airway about one eighth of an inch in diameter through the smallest part. The large end, about half an inch in diameter, was held in the mouth and when air was blown down the tube on to a flame a jet of flame was produced. The flame could be from a spirit lamp or from a handful of rushes tied together to form a torch. The blowpipe could be used to make the blown joints or blowpipe joints.

BOSH

Name for a sink in South Wales. The OED tells that in mining, a bosh is a trough for cooling bloomary tools, ingots, etc; a bloomary is the first forge in iron works. An old catalogue of blacksmith's equipment illustrates a water bosh and tue iron complete as used in conjunction with bellows and hearth back. The Welsh bosh, meaning kitchen sink, took its name from the main local industry.

BOSS

A protuberance on a vessel, or a fitting for attaching to a vessel, for making easy the connection of pipe or fitting to the vessel.

There is also the tap boss, a short fitting with plain end for soldering to lead pipe and a female end to receive the male thread of a bib tap, and, if the tap boss is too short, there is a tap extension boss with male thread one end and female thread the other; this is sometimes called a 'tap extension piece', a self-explanatory phrase.

Bossing

Outside plumbing circles, to boss is to fashion in relief, or

beat *out* into a raised ornament, say; usually called emboss-
ing. When a plumber bosses sheet lead he beats it, or dresses
it, or even just works it, and not necessarily *out*. He may
make a box with external and/or internal corners—that is
bossing. The art of bossing is the moving, by beating in
appropriate manner, of the thickness of the lead to where it
is required; when the work is finished, the lead should be
of an even thickness throughout.

Whilst the English or Welsh plumber bosses his lead, the
Scottish plumber will work it; in the south one bosses a
corner, like that of a box, say. In the north one knocks up
or works a corner.

BOSS AND FLANGE CRAN

A bib tap of a design found mostly in Scotland; plumbers in
the south of England are not familiar with it. It has, on the
back of the body, an integral flange of about three and a half
inches diameter which acts as a fixing plate to a wooden
casing at the sink. In Scotland this casing is called a 'cran
box' and it acts as a shelf for soap, etc, as well as concealing
the pipes at the sink. The fixing of the cran box is, of course,
the carpenter's job.

The boss is the short brass fitting with female thread for
soldering to lead water pipe.

BOSSING STICK

A round dresser (qv). In the southern half of England
and in Wales, any dresser other than the flat one is a 'stick'
—the round dresser becomes a bossing stick—a stick for
bossing.

BOX HEAD

A rainwater head of any shape in parts of mid-Wales.
Probably from the old method of receiving roof water in
wooden gutters which passed it to a wooden box at the top
of a wooden rainwater pipe. All that was the work of the

carpenter and the plumber would be called in if the box had to be lined with lead, and in that case the gutter and the rainwater pipe would be of lead, too, if cost allowed.

BRIG-AND-NAIL

An old Scottish fitting for connecting a sink grating to a lead trap. The term, when translated to English, means 'bridge and nail'; the bridge was a brass bar of a length little more than the diameter of the lead trap and with a threaded centre hole. Two holes were made in the wall of the trap, one opposite the other, a short distance from the top of the inlet of the trap. The bridge was slipped in across the pipe so that each end of the bridge protruded only slightly through the holes. Solder dots were then made over the ends and the holes to secure the bridge. A lead flange could be soldered on to the inlet or the inlet flanged out with a mallet. The nail, a long, threaded bolt, passed down through the grating, screwed into the centre hole of the bar and pulled up the trap flange to the sink on the same principle as the better-known skeleton waste.

The brig-and-nail was also known as the bar-and-nail.

BROG

Cobblers and Englishmen have awls and bradawls. But the Scots have brogs—all words for the same piercing tool. The brog is for making a hole in wood when putting in a wood-screw. As a verb it is used as 'brog a hole' or 'brog a piece of wood'.

B Th U

An epitaph for the British Thermal Unit, long known as the amount of heat required to raise 1lb of water through 1F° or, conversely, the heat given off by 1lb of water when cooled through 1F°.

The first sign in the deterioration of the exactness of the definition began probably in the 1940s when 1F°, in many quarters, became 1°F; the difference being that 1F° is simply any one degree on the Fahrenheit scale, while 1°F is a specific temperature.

Then, into the 1950s, when plumbing was said to be becoming a highly technical and precise science, the British Thermal Unit, previously known as B Th U, became a BTU.

BTU

As well as now meaning B Th U, the BTU has always been known as a Board of Trade Unit of energy, 1kW per hour —1kW per hour is the equivalent of 3,412 B Th U.

btu

Bandied about by a central-heating-conscious public and technicians alike, the BTU was badly bruised and misused until about the mid-1960s when advertising and technical articles reduced it to btu.

BUCHAN TRAP

For so long has the Buchan trap been recognised in Scotland as *the* disconnecting trap (see DISCONNECTOR) that 'Buchan trap' has become a part of Scottish plumbing language and synonymous with disconnector or disconnecting trap. Although this trap is always tagged as Scottish when written about outside the land of the kilt, there is evidence, from good authorities, that it was used extensively in the south, in cities like Leicester, Cambridge, Bath and Bristol.

'Buckin' trap', a local corruption, is quite common as is a similar and more crude descriptive adjective when a choked trap turns out to be a stinking pick-and-shovel job; Scottish children have for years been expert at engineering 4in-plus tin-cans down 4in fresh air inlets into traps.

In Scottish dialect 'buck', with reference to water, means to gush out, to gurgle when poured from a straight-necked bottle, or to gulp in swallowing; those meanings describe well the Buchan trap in action. In early stages of research it was easy to conclude that 'Buchan' was a corruption of 'buckin' '; in fact, the trap was patented by W. P. Buchan in 1875 and the similarity which buckin' has with 'Buchan' is coincidental.

The idea of ventilating drains by means of a fresh air opening in a trap was probably first conceived by a Glasgow plumber, Wallace by name, in the employment of a Mr Lockhart. In 1853, Wallace was sent to cure a smell in a gentleman's house at Garnkirk, near Glasgow, and, being a thinking plumber, he hit upon the idea of cutting a hole on the house side of the trap and from the hole he carried a pipe to the roof. So successful was Wallace in ridding the house of stench that it soon became known in the trade that Lockhart's men were doing this thing.

Buchan was a young apprentice at the time of Wallace's experiments but he, too, was said to be 'a thinking lad' and there is little doubt that for many years he nursed the problem of ventilating drains, having gleaned the seed of the idea from Wallace. The trap was Buchan's answer to the controversy about drains which arose from the illness of the Prince of Wales in 1871. It was praised by Hellyer and Davies.

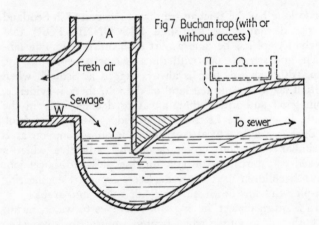

Fig 7 Buchan trap (with or without access)

The main features of the trap were :

(1) 'A' was a fresh air inlet with cutaway in the pattern to allow for easy in-flow.

(2) The surface of water exposed at 'Y' was no more than the bore of the drain. Owing to this and to the drop from the square edge at 'W' which the water got when cascading from the branch drain, the faeces broke up and were carried away. The square edge at 'W' was more self-cleansing than a round edge.

(3) The seal, or waterlock, at 'Z' was no more than $1\frac{1}{2}$in.

(4) The access on the outlet could be used for cleansing, cleaning off the inside of the cement joint when fitting, or it could be extended as a sewer-gas outlet.

The trap could be had in one piece or in two pieces. The former is illustrated.

There were a great many attempts to copy the Buchan trap and many legal actions for infringement of copyright. That the infringers won in the long run is apparent, because the present-day Buchan trap bears only small resemblance to the original. Indeed, the one laid down by the British Standards Institution completely misses the points of the main features in the original. The surface water at 'Y' is greater than the bore of the drain; there is no cut-away at the fresh air inlet; there is a rounded edge with splayed drop at 'W' instead of the vertical drop, and for some years the seal has been specified as $2\frac{1}{2}$in instead of the original $1\frac{1}{2}$in.

For some years the Buchan trap has been said to be a nuisance because of constant blockage, but it must be pointed out that it is not the original that is being condemned.

BUCKET HEAD

A rainwater head. The term has been heard in Cornwall, Somerset and Co Durham, but a number of old plumbers connect the term with the heads of old jack pumps.

The bucket head, meaning rainwater head, is very often bow-fronted in the form of a bucket and it is likely that the origin of the term is just as simple as that, although a bucket head is not supposed to hold water, as a bucket does. Another possible origin for the name could be in the nineteenth-century term, soil pipe head (qv).

D

CAME

A grooved bar of lead used for framing the glass in latticed windows. Although the making of such windows, usually called leaded lights or leaded windows, is no longer done by plumbers, many old workmen still tell of that branch of the craft.

Few of the old plumbers speak of 'cames', the terms varying from cams, calmes, to carms.

CANDLE

In Scotland, the male end of two prepared ends of lead pipe when jointing is called the candle, candling, or candled end. 'Candle' and 'candling' are also verbs for the forming of a candle; to candle a pipe, the end is rasped to a shape roughly resembling a candle.

CAP AND LINING

A union coupling, ie a screwed coupling which facilitates dismantling, the cap being a coupling nut for connecting to a threaded pipe or fitting, and the lining being a brass or copper tube, bent or straight, for soldering to lead pipe. The lining has an integral collar at one end for engagement by the cap when the connection is made.

This is a term used in England, particularly in the southern half, and in Wales. The parts of the term are difficult to understand as the word 'cap' suggests something for capping off or plugging off; this cap is merely a nut, usually brass with an internal shoulder at one end. The lining does not line anything as the name might suggest; it merely acts as a connecting pipe or tail between lead pipe and another pipe or fitting.

Cap and Tail

This variation has been heard in Llandiloes, Wales.

Coupling

The Scottish term; very few Scottish plumbers are likely

to know 'cap and lining' unless they have travelled as plumbers in England. The word 'coupling' must be qualified, like bath coupling—a coupling for connecting a lead pipe to a bath tap—basin coupling, waste coupling, etc. The nut part of the coupling (ie, the cap) is the coupling ring and the connecting pipe or tail is known as the coupling tail.

Union Coupling

Term used in BS 4118, the two components of the union coupling being a 'coupling nut' and a 'tail pipe'.

CAULK

In general terms, to stop up seams or crevices by driving in soft rope, cord, or similar material. In plumbing, to pack a spigot and faucet, or spigot and socket, joint with gaskin and, having filled the remainder of the space with lead, to drive home the lead with caulking iron. Under certain circumstances, and in certain areas, the spigot and faucet joint is caulked with gaskin and the remainder of the space filled with cement or even just putty, but such joints, although caulked with gaskin, are not considered as caulked joints. The word 'caulk' seems to infer the making of a joint with gaskin *and* lead, whether the lead be poured in when molten or in the form of lead wool or lead rope.

The fifteenth-century spelling was calke or caulke; in the sixteenth century, calk or caulk. The OED says the 'calk' spelling is no longer in use (see also US CALK). However, regardless of the spelling, the pronounciation has always been *kok*; many plumbers try to include the silent 'l'.

In Scotland, in particular, the caulking of the lead is called batting. (See also BAT)

Cork

A variation of caulk heard in Cornwall.

Set up

In London area, pertaining to caulking lead in spigot and socket joints.

Stave

Heard near Hawick and in Lanarkshire, Scotland, with reference to caulking the lead in a spigot and faucet joint. Chambers gives the verb 'stave' the meaning : to push, drive; beat against; thump vigorously; as a noun it is a heavy blow. Any of those meanings could well be used to describe caulking. The OED gives this meaning of 'stave' as Scottish and US.

CAULKER

A plumber who specialises in laying cast iron water mains, gas mains, and drains, and in the making of the necessary caulked joints. Frequently referred to as a 'pipe jointer'.

CAULKING TOOL

A tool of the cold chisel type but with blunt face and with a double set in the length to allow fingers clearance when caulking joints in spigot and faucet joints. The thickness of faces varies and a plumber usually has a number of caulking tools of varying thicknesses as well as with varying sets; certain types have sets which allow caulking to be carried out on the back of vertical pipe and on the underside of horizontal pipe, say, in a trench or duct.

Caulking Iron

While 'caulking tool' is common in England and Wales, 'caulking iron' is just as common in the extreme North of England and in Scotland.

Batting Iron

Scottish term for a caulking tool with a thick face, usually used for finishing off a neat caulked joint. (See also BATTING IRON)

Setting-up Tool

Heard in the London area as the tool for setting up (see CAULK) a caulked joint.

Staver

Or staving iron. Terms for caulking tool used in Scotland (see CAULK).

CENTRE GUTTER

A gutter with a flat sole and two sides angled to the slopes of adjoining pitched roofs. The illustration shows a lead centre gutter.

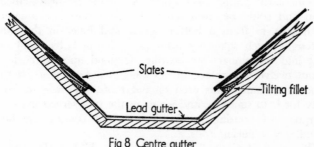

Fig 8 Centre gutter

Valley Gutter

The centre gutter is known as a valley gutter, as often as not, in England and Wales, but this term can cause some confusion as there is another type of valley gutter (see FLANK).

CESSPOOL

The Scottish drip box (qv) is known in England and Wales as a cesspool. That a box for receiving rainwater should be called a cesspool is rather puzzling as one usually associates cesspools with the accumulation and storage of foul water for collection at a later date (see also SINK). A cesspool on a roof may collect grit and leaves which have to be cleaned out periodically but, as the cesspool has a bottom outlet, it has no other likeness to the foul-water type.

CHAFFER

Chambers gives 'choffer' as an alternative spelling and meaning chafing-pan. The OED gives chafing-pan as a vessel to hold burning fuel for heating anything placed upon it; a portable grate. The Scottish chaffer (or chaffer-pan) is usually pronounced 'choffer' and can be something as simple as a perforated bucket or drum for holding burning fuel, and such a portable grate was often used by plumbers to heat the melting-pot or even just to boil up the billy-cans.

The early plumber's chaffer was made of heavy sheet iron and could be from ten to fifteen inches in diameter, with bars to form a bottom grate and holes in the side through which soldering irons or soldering bolts could be put into the embers for heating. It stood upon three legs about twelve to eighteen inches long. A necessary accessory, when the chaffer was used on flat roofs, say, was an iron tray for it to stand in, and water in the tray served the two purposes of dousing hot cinders and of cooling the hot handles of soldering irons.

The verb chafe, to heat, is from the old French *chafer*. (See also DEVIL)

CHALK LINE

When cutting strips from large sheets of copper, lead or zinc, a chalk line is used for 'striking' a true mark to which the plumber can cut. The line is rubbed with a knob of chalk and (the required measurements being marked at two points on the sheet) the plumber and his assistant then hold the line very taut from mark to mark. One of them plucks, or strikes, the line, and an accurate chalk mark is made on the sheet.

To strike a vertical line on a wall, the line is held at the top of the wall and the plumb bob allowed to swing until stationary. Without deviating from the vertical, the mate pulls the bob down until the line is taut, then strikes with his free hand. (See also PLUMB BOB)

CHICKET

Name for a dormer window in Cornwall and North Devon. Correct spelling uncertain and could not be verified. A few plumbers did not recognise the word 'dormer'.

Dormers have cheeks and there may be a connection here with 'chick' and cheek. But there could also be some connection between chicket and the French *chicot* meaning stump of a tree, an old tooth, or the remains of a broken tooth; seen against the skyline the dormer window gives one the impression of a tooth, or a broken tooth.

CHIMNEY

In the matter of flashings the plumber has an interest in chimney stacks, loosely called either 'chimneys' or 'stacks'. Throughout Britain the old generation of plumbers, in common with old people in general, use variations like chimblay or chimbley, chimbla, chimley or chimla.

Lum

As well as the above variations, a chimney is a lum or lumb in Scotland. It may mean a chimney corner, a chimney-stack, or a chimney pot; chimney sweeps sweep lums; a top-hat is a lum-hat, sometimes called a tile-hat—chimney pots are sometimes made of tile.

CHIMNEY BACK

In North Wales, the gutter at the back of a chimney on a pitched roof.

In other parts of Britain it is a chimney gutter or a back gutter.

BS 2717 points out that while a chimney gutter is a back gutter, a back gutter is not necessarily a chimney gutter. It says :

| Chimney gutter | A gutter formed at the back of a chimney stack penetrating through a pitched roof. |
| Back gutter | A gutter formed at the back of a chimney or other penetration through a pitched roof. |

CISTERN HEAD

A usual name nowadays for a rainwater head in Wiltshire, Cornwall and parts of Wales. But it was not always so, for Davies and Hellyer, both well-known London plumbers over the turn of the century, used the term in their writings. Davies refers to it as the square or cistern head; in local areas it is now applied to heads of any shape.

Most of the early heads were basically of a rectangular cross section, like a cistern, however intricate the ornamentation.

CLACK VALVE

A non-return valve (see REFLUX VALVE). A clack is a sudden, sharp, dry sound as of two flat pieces of wood striking each other, or anything which makes this noise, according to the OED. The clack valve was used in the earliest form of pumps for raising water from wells. The clack is the working part of the valve, a weighted flap which opens by the upward motion of the water and closes to retain a quantity of water until the next stroke of the pump. The non-return valve used nowadays in hot water systems and in any circumstances where there is likely to be an undesirable backflow, works on the same clack principle and is still called a clack valve in many parts of Britain.

Retaining valve

A term used in an MAC catalogue (1927) for a clack valve.

CLINK

Although welt (qv) may be heard in Scotland and the North of England, clink or clench are more usual.

CLOSET CISTERN

Most plumbers and laymen know a closet cistern when they see it : it is the cistern fixed over a closet pan from which a quantity of cleansing water is discharged into the pan when a chain is pulled or when a lever or a knob is pushed.

Flush Box

Heard in South Wales. This probably derives from the days when closet cisterns were often made of wood and lined with copper or lead.

WC Flushing Cistern

BS 4118 does not admit the existence of the term 'closet cistern' but gives 'WC flushing cistern'. As a WC, according to this British Standard, is a compartment and not a pan, it must be understood that the WC flushing cistern is not for flushing out the compartment; it is merely fixed *in* the compartment for the purpose of flushing out the pan.

COCK

The earliest record of the word is in the Old English form, before 1150; cocc, coc, kok, the word probably echoic, says the OED; an imitation of the cluck of the common domestic fowl.

By 1481 it was a spout with an appliance for controlling the flow of liquid through it, and it retains this meaning today. It did not take on the meaning of penis until 1730.

Little children who are curious about the penis are often told that it is a 'bird' and that must make them still more curious about birds. The penis is probably referred to as a cock because it is 'a spout with an appliance for controlling the flow of liquid' and not from any association with the male domestic fowl.

A pipe on a wall, which should be vertical and is not quite so, is said to be cocked, or even 'all to cock', but that meaning is probably taken from the angle of the hammer of a gun when cocked; the hammer of a gun, the cock, is so called because of its original shape—resembling a cock's head. (See also VALVE)

COCK HIGH

A phrase often used to indicate a suitable height from the floor to the top front edge of a fireclay sink or wash basin.

24in from the floor to the underside of the sink has long been the recognised height of a fireclay sink unless the plumber is asked to fit it otherwise. Although 'cock high' is a jocular rule of thumb suggestion, it does call traditional sink-heights to question. In the belief that ideally a housewife should be able to place her hands on the bottom of the sink without bending her back, this plumbing-author has frequently held a fireclay sink up on a wall so that a housewife could decide on a position which suited her. Just as 'cock high' varies from man to man, so sink heights should vary from woman to woman; there would be fewer aches and less anguish.

COCKNEY FLASHING

In a few cases this phrase was given as meaning herringbone flashing when used in conjunction with soakers (qv). Throughout England and Wales, it is simply herringbone flashing. (See STEP FLASHING.) 'Cockney flashing' was uttered with contempt by an elderly Welsh plumber who considered it rubbish compared with the single step method.

There has been a strong indication that the herringbone flashing originated in London and thereabouts.

COLLAR

In London and the south east the faucet or socket of cast iron pipe is often referred to as the collar of the pipe, but

BS 4118 says that a collar is a fitting in the form of a sleeve for jointing the spigots of two pipes in the same alignment; the BS type of collar is better known to plumbers as a double collar or loose collar, but those are now non-preferred terms, while other well-known names, sliding collar and slip collar, are equally deprecated.

In Somerset, a gentleman of the Water Board called a cast iron pipe-socket (or collar) a double collar. When asked about loose collars he said : 'Oh, those too are double collars or slip collars.' (See also COUPLER)

CONDUCTOR

BS 4118 gives 'rainwater conductor' as a non-preferred term for rainwater pipe, and 'soil conductor' as non-preferred under soil pipe. 'Soil conductor' is not a common term; it has not been heard in course of this research. The word 'conductor' need not be qualified in Scotland or northern parts of England as it is understood to mean a rainwater conductor. The term is not usual in the southern half of England. (See also US LEADER)

CONNECTOR

A device for connecting two pipes, one or both being fixed, or for connecting a pipe to a fitting or to an appliance. There are flush pipe connectors for connecting flush pipes to WC pans; group connectors for connecting waste pipes to gulleys; urinal connectors for connecting urinal gratings to cast iron waste pipes; WC connectors for connecting soil pipes to WC pans. However, when the word 'connector' is used by itself, it usually means what BS 4118 describes under 'longscrew' (qv).

When given the name 'connector', the majority of plumbers interviewed in England and Wales gave a similar description along with alternative terms (see also RUNNER).

The National Price List includes longscrews and 'connector and backnut'; this longscrew is not a connector. Also, as a connector comprises the three components, longscrew, backnut and socket, the 'backnut' is unnecessary in the Price List's 'connector and backnut' (see BACKNUT).

When connecting two pipes with a connector, there must be sufficient play in one of the pipes to allow for the screwing on of the connector. Once it is screwed on to one pipe, it can be held in alignment with the other and the socket screwed to it, followed by a grommet (qv) and then the backnut which squeezes the grommet against the face of the socket.

Double Connector

A short piece of pipe with a long thread on *both* ends and, subsequently, a backnut and socket on each. It is used to connect two pipes fixed rigidly in the same alignment. The double connector, being of an exact length to bridge the gap between the two pipes, is held in position and the two sockets screwed on to their corresponding pipes, then the grommets and backnuts are tightened.

COPPER BIT

A copper bolt (qv). Any tool with a biting edge, like a carpenter's bit, say, can be called a bit. When soldering with a hot copper bit, the biting edge, or point, bites solder from the stick, thus the term. The name is better known in the southern half of England than in other parts.

COPPER BOLT

The copper bolt has a head of copper on an iron stem with wooden handle, and is used when soldering with fine stick solder. When plumbers had to heat the bolt in a fire, the handle was loose fitting so that it could be slipped off to prevent charring while the bolt was heating. The pointed end of the bolt is 'faced', or tinned, by filing clean while hot and quickly dipping the point into flux and touching with solder. A good face on a bolt makes for easy fusion of the solder.

The terms copper bolt, soldering bolt and soldering iron

are commonly heard in Scotland, northern parts of England and in Wales (see also COPPER BIT) and are understood to mean the straight bolt as shown (cf HATCHET BOLT)

Fig 9 Copper/soldering bolt/iron

Davies says that the word bolt originates from the time when ships were of oak and copper bolts were used to bolt together the oak timbers; the gallic acid contained in oak rapidly corrodes iron-work, thus preventing the use of iron bolts in such ships. The copper bolts were sought after by plumbers who made them to an appropriate shape for soldering.

COUPLER

Still used in many parts of Britain for a pipe fitting for jointing two pipes.

In the case of couplers for copper tube, the 'coupler' is listed in catalogues as a straight coupling which infers that there are bent couplings or couplers, but the fittings which might be called bent couplings are listed merely as bends or elbows. The term 'bent coupler' has been used in Scotland for many years with reference to what is now an elbow.

For mild steel tubes, the coupler is a short barrel-like fitting threaded on its inner surface, and is used for coupling together two pipes. Such couplers are always straight—in literal terms an elbow or knee could be called a bent coupler. The word 'coupler' as used for a mild steel tube fitting is on its way out, as BS prefers 'socket'.

Coupler, socket, collar and sleeve are words for this fitting which might be heard in everyday plumber language almost anywhere in Britain.

CRAMP-AND-RATCHET

An early piece of equipment—there may still be a few in existence—for boring and tapping water mains and gas mains. It has been superseded by more sophisticated machines which allow boring and tapping of the mains without turning off the supplies. The cramp-and-ratchet consists of a kind of jig with two semicircular hooks for gripping the underside of the pipe, the drill or tap being fitted into the frame of the jig on the top side and rotated with the aid of a detachable ratchet. The term was heard in Berwick-upon-Tweed.

Shangie

An informant in Berwick-upon-Tweed gave this Scottish word as used by employees of Newcastle and Gateshead Water Company for 'cramp-and-ratchet'. Chambers gives 'shangie, shanjie, shangy, v. to enclose in a cleft piece of wood.—n. a cleft stick for a dog's tail . . .' The connection between that definition and the boring gear would appear to be the clefting of the pipe with two semicircular hooks or clefts. (See also SHANGIE)

CRAN

At least one Scottish apprentice plumber has associated the cran, or crane, with Crane Ltd, makers of valves and fittings, but it has been established that the name Crane was not borrowed from Scottish plumbing. The Crane organisation was started in the United States of America by Richard Teller Crane in 1855 and it was not until 1918 that Crane Ltd was established in Britain.

When asked about a 'cran', Sassenach plumbers have thought that it had something to do with fish; cran, probably from the Gaelic *crann* meaning a lot, does indeed mean a measure of herring, about seven hundred and fifty fish.

But in Scottish dialect, cran is the equivalent of crane, and can mean heron, swift, crane, an iron instrument laid across the fire to support a pot or kettle, or a piece of machinery for lifting heavy gear.

Chambers gives all those meanings plus one of interest to plumbers, ie a bent tap. To Scottish plumbers a bent tap is a cran; to English-speaking Scots it is a crane. But under 'crane' Chambers gives 'the tap of a gaslight or of a barrel'. That suggests there is some slight difference between a cran and a crane. But regardless of spelling, any of those things mentioned is still, in the dialect, a cran.

The OED gives crane, from the Old English cran, and amongst other meanings, a bent tube for drawing liquor out of a bottle; a siphon, 1634. So cran does not seem as Scottish as one would suppose but rather one of those words rejected by the English and retained by the Scots.

BS 4118 gives crane (Scottish) as a non-preferred term under 'tap', which is defined as 'a valve with a free outlet used as a draw-off or delivery point'. The main feature of a cran or crane is not mentioned, namely the bent neck or bent outlet. The BS definition of a bib tap (qv) fits the description of a crane : 'A tap with a horizontal inlet and a nozzle bent to discharge in a downward direction.'

In Scotland there are bib crans, basin crans, bath crans and, of course, the misnomer 'ball cran'.

CREEPERS

Probably obsolete, but in the early twentieth century they were short mandrels, three to five inches long, for driving bobbins and followers (see BEADS) through a pipe when bending. Creepers were often referred to as dogs.

CROCK

Usually associated with earthenware pots, and machines, men and beasts which are frail and past their best. However, in South Cornwall, elderly plumbers referred to the melting pot as 'the crock'. The term does not seem to have been confined to South Cornwall; the OED gives it as a 'metal pot' in the south west of England, 1475.

CROOKED THREAD

A pipe thread which is slightly out of alignment. When mild steel pipes are being put into old properties, say, in which the angles of walls are slightly more or slightly less than right angles, a 90° elbow, say, may cause the pipe to run out of line with the wall. In such circumstances, the old method of making a crooked thread is often adopted. It is done by easing off the guide of the dies, so that a lopsided pull, whilst making the thread, will result in the teeth of the dies biting deeper on one side of the pipe than on the other. When the elbow is screwed on to the thread the angle created by the elbow will be slightly more than 90° or slightly less, depending on the position of the tightened elbow.

CROW'S FOOT

Also known as a basin spanner or basin wrench. It has a pair of jaws at each end turned at right angles to the stem and facing in opposite directions. The jaws at one end may be suitable for the under nut on basin supplies, and the jaws at the other for the nuts at bath taps.

As this 'basin spanner' is meant for work at baths as well as basins, 'crow's foot' is probably a better term.

CRY OF TIN

A term used by an elderly plumber in Llandello, with reference to the crackle heard from a strip of tin or from a tin pipe when it is bent near the ear. He likened the crackle to the cry of a baby, thus the cry of tin.

CUNDY

Scottish. Given by Chambers as 'a covered drain; a concealed hole, an apartment; the hole covered by a grating for receiving dirty water for the common sewer; a small drain crossing a roadway'.

Cundies still in use in the 1930s and probably later were constructed with flat stones in two shapes. One type, with triangular cross-section, had a flat stone base with other stones straddling the base to form the two sides of the triangle. The cross-section of the other form was rectangular with base stone, two vertical sides supporting the top stone. Cundies were used to take surface water and waste water from sinks and gullies.

The Old French form of the word, 'conduit', survives in England and is used to mean an artificial channel for carrying water. The word is pronounced 'cundit'.

CUP HEAD

Heard in mid-Wales and Co Durham for a rainwater head. The term is reasonable enough; if there are pot heads and bucket heads, why not cup heads? Many of the small cast iron rainwater heads are so dainty and cup-like that they could not possibly be called anything else.

CUP JOINT

A term often used to describe the taft joint (qv) but in preparing the cup joint, the lead is merely cupped and not tafted or saucered over. The resulting cup, or socket, means that very little solder is used in the making of the joint; many quite competent plumbers scamp their work by using this joint in concealed places to save solder and to turn out a quick job. Because of the increasing number of modern plumbers who are incapable of making a proper round joint, many water authorities permit the use of the cup joint on underground lead water pipes. The term was heard in Cornwall and mid-Wales.

CURB ROOF

A mansard roof. 'Curb' is usually associated with a raised border or edge and is used as a variant of kerb. But another meaning of the verb 'curb' is to bend or curve; the rafters of a curb roof are bent downward at the lower ends, the curb being at the intersection of pitches.

E

DEVIL

The English term for the Scottish chaffer and sometimes called fire-devil. The devil, as described by Davies, was a little more sophisticated than the chaffer illustrated by Buchan; indeed, the devil of that time had an arched piece of iron over the top from which to hang the melting-pot, similar to the devil of recent years.

DISCONNECTOR

From the English Midlands northward and throughout Scotland the interceptor is better known as the disconnector or disconnecting trap. While the English disconnector is usually contained in a manhole, the Scottish one (see BUCHAN TRAP) is usually buried and served with fresh air through a vent pipe which finishes at pavement level. It is so called because it disconnects sewer gas from the house drain.

DOG EAR

As copper and zinc cannot be worked like sheet lead to form corners, the external corners necessary in sheet copper and sheet zinc work are formed by folding the metal. This is known as dog earing in England and Wales. The dog ear is sometimes made in sheet lead as well if the worker is incapable of bossing lead, and also by quite competent plumbers to save time on a job.

DOG TOOTH

Or dog-tooth flashing. Term used by plumber in North Riding for herringbone flashing.

DOLLY

Neither Buchan, Davies nor Hellyer used this as a plumbing term, but Davies, in a short glossary of obsolete and trade words, describes a dolly as a Sam shop or illegal pawn shop, the name being taken from the sign of a black doll, once the sign of a rag shop. The black doll Sam could have become Uncle Sam which, further curtailed, became the present day 'uncle's', meaning a pawn shop.

However, in plumbing circles, 'dolly' has been used as one might use thingummybob or what's-its-name when groping for a word. In such a way, dolly has come to have a particular meaning within the limits of small areas and even within individual workshops. But it is certain that the dolly is something with which to deliver a dull, heavy blow, or which receives such a blow.

Mallet

In one Scottish workshop, at least, in Lanarkshire during the 1930s, a dolly was a kind of mallet with a large spherical head of *lignum vitae* for thumping bobbins into the end of $3\frac{1}{2}$in lead soil pipe to form a socket into which a closet outlet would fit.

Bobbin, Tampin

As recipients of dull, heavy blows, both these tools were referred to as dollies by elderly plumbers in South Cornwall.

Dummy

With its head of lead this is an ideal tool for delivering that dull, heavy blow. 'Dolly' was given as an alternative of 'dummy' in Kent, Surrey and the London area. Plumbers in various parts of the country, who were not familiar with dolly, guessed that it could be a dummy.

DOOK, DOUK

A wooden wedge driven into a wall to hold nail or screw. A word used by layman and tradesman alike in Scotland; in England and Wales, such a piece of wood is called a plug.

To make a dook, take a block of wood slightly larger than the hole to be plugged and with the grain running length-wise. With knife or hatchet slice off two diagonally opposite corners for about three parts of the length so that one end of the dook retains rectangular section and the opposite end is little more than a narrow edge. The resulting propeller-like, or duck-tail-like, shape of the inserted end causes the dook to twist as driven into the hole, and twixt twisting and driving makes a solid fixing. This traditional method of plugging a wall has an advantage over patent plastic plugs in that the dook can be of a size to take more than one nail or screw.

Dook, as a verb, means to cut a hole in a wall for a dook; one uses a dooking-iron when dooking a hole for a dook.

In Scottish dialect, dook and douk are variations of the English duck, but not the feathered duck; that is a deuk. So it is unlikely that this wooden wedge takes its name from its shape. To dook is to dip or bob down quickly; in Scottish coal industry, an inclined road or dip in the mine is a dook. On a smaller scale, cutting a hole in a wall can, with some imagination, be compared with cutting a dook in a mine. The only incongruity is that in one case the dook is the plug, and in the other it is the hole.

Tampin

As the tampin is a type of plug (see TAMPION) it is logical that it should be known as a dook or a plug in various areas of Northumberland, Cumberland, Dumfriesshire, Peeblesshire, Roxburghshire and Berwickshire; the nearer the Anglo-Scottish border, the more frequent is the use of dook or of plug.

DOOKING IRON

Although the Scottish dook is known in England as a plug or wall plug, the dooking iron is not similar to the plugging chisel of the south; the plugging chisel is more like the Scottish raggling iron.

The best dooking irons are made on the anvil by a blacksmith and out of old files, because of the hardness. It is different from an ordinary cold chisel in that the blade of the dooking iron is drawn out wider than the shank; as the blade is the widest part, a hole can be cut to any required depth in brickwork or masonry without sticking or binding. An ordinary cold chisel with the blade the same width as shank starts binding before any depth of hole is made. The plumber and also the carpenter in Scotland have an assortment of dooking irons suitable for making holes of various sizes.

DORMER

A vertical window in a pitched roof. It may have front elevation of rectangular shape, arced or eyebrowed, or semicircular. The sides of a dormer, which the plumber may be called to cover with lead, zinc or copper, are the dormer cheeks (see also HAFFIT). The structure may be set on the angle of the roof, partially recessed or wholly recessed as illustrated in outlines.

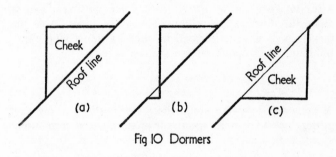

Fig 10 Dormers

Storm Window

The wholly recessed dormer is called, by Buchan, a storm window. This type has a platform and cheeks in front of the window itself; those may be covered with any of the plumber's non-ferrous metals. (Fig 10c.)

'Dormer' is descended from the Latin *dormitorium,* a sleeping room. In the sixteenth century it would be the window of a dormitory. (See also LUKIM, CHICKET and GARRET WINDOW)

DOUBLING

In Scotland tilting fillet (qv) is frequently called doubling or doobling, probably because sheet lead goes up and over it, ie doubles over it. (See Figs 8 and 16)

DOWN-COMER

A rainwater pipe. Although BS 4118 gives this as a non-preferred term, many regions in England and Wales are not familiar with it and it is not in the Scottish plumber's vocabulary. It has been heard in Co Durham and random places in central England.

DOWN PIPE

May be heard practically anywhere in Britain for a rainwater pipe. For very many years, plumbers have used the witticism 'putting up down pipes, not putting down up pipes' to describe putting up rainwater pipes. Any pipe down which fluid falls could be called a down pipe, but in plumbing it is specifically a rainwater pipe.

DOWN-SPOUT

A rainwater pipe. Although there are localities where eaves gutters are known as spouts or spouting, the term down-spout is not necessarily used in those places. In parts of Lancashire where the eaves gutters are eaves gutters, the rainwater pipe is a down-spout. Down-spout has also been heard in Cumberland and across the border into Roxburghshire. (See also US LEADER)

DRAIN PIPE

To plumbers a drain pipe is a drain, usually underground, comprising lengths of *drain pipe* which may be of fireclay, cast iron, pitch fibre, or PVC. To the layman, 'drain pipe' can mean rainwater pipe, waste pipe, soil pipe, or indeed any pipe on the face of a building.

DRESSER

In plumbing as in other trades, to dress means to make smooth or to make presentable. A carpenter uses a plane to dress a piece of rough timber; the plumber uses his dresser to dress sheet lead, to make it flat and smooth.

Dressers are shown in a London toolmaker's catalogue as being of boxwood, beech, or hornbeam. A fastidious plumber is careful that his lead is not marred by unnecessary dresser marks, and he selects his dressers and uses them at the right time and on the proper occasion with an eye to the weight and to the sharpness of the side edges. Dressers are handed; that is, they are designed for use on the right or the left hand of a gutter, say.

Ask a plumber in the southern half of England for a dresser and he is likely to produce a dresser as he knows it—with a flat beating surface. But from the Midlands northward and throughout Scotland, the plumber would want to know what kind of dresser—the dresser in the south is a 'flat dresser' in the north.

In *The Practical Plumber and Sanitary Engineer* J. Malpass writes, in a section on tools, of flat dressers, oval

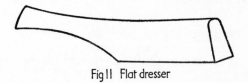

Fig 11 Flat dresser

dressers, and round dressers, so one can deduce that he does not come from much further south than the Midlands. (See also SPOON DRESSER, ROUND DRESSER, SETTING-IN DRESSER)

DRIFT

A steel cylindrical tool for driving into the end of a copper tube when flaring out the end as required when using certain compression-type fittings. Each size of tube has its corresponding drift and each make of fitting should be used in conjunction with its particular make of drift as the taper and style vary from maker to maker of fittings.

In the everyday, accepted sense, the word suggests a slow, aimless movement, but it really comes from the medieval *drifan*, to drive. The OED gives many meanings varying from 'natural or unconscious course . . .' to 'the conscious direction of action . . .' and also 'that which is driven'. The plumber's drift is driven with hammer; a spit on the drift makes manipulation easier.

Perhaps because of its general and accepted sense, plumbers subconsciously reject a drift as something that is driven, and consequently throughout Britain other unimaginative terms are substituted : dolly (qv), expander, expanding tool, flaring tool, opener, swaging tool and widener.

DRIP BOX

A box-shaped sump formed in roof construction and ultimately lined with sheet metal, at the lowest end of a gutter, for collecting rainwater which then passes to a rainwater pipe. Drip box, sometimes dripping box, is a Scottish term. (See also CESSPOOL)

DRUNKEN THREAD

May be heard frequently from one end of the country to the other meaning a crooked thread.

DUCK FOOT BEND

A drain bend with a foot forming a reinforced base with which to withstand the weight of a vertical pipe stack. Although this term has been used for about one hundred years, British Standards have decided, and makers have followed suit, that 'rest bend' is the preferred term.

Hellyer, Davies and Wright Clarke all used the term 'duck foot bend', and Buchan and many plumbers since his time called it a 'boot', or 'boot bend'.

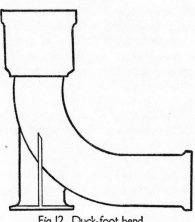

Fig 12 Duck-foot bend

DUCK'S MOUTH

Heard in Bristol as an alternative to bird mouth.

DUMMY

A tool which the plumber usually makes for himself, consisting of a cane or a piece of ½in mild steel gas pipe with a knob of solder about the size of a hen's egg on one end. It is used for knocking out, from the inside, kinks in large diameter lead soil pipes when bending same and for using

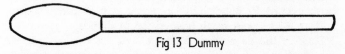

Fig 13 Dummy

in conjunction with the round dresser in bossing corners in sheet lead.

If the ferrule of a broken drain rod is intact, a good dummy can be made by tinning the ferrule and shaping the knob as if wiping a solder joint.

Hand dummies are usually from twelve to eighteen inches long, and long dummies about thirty-six inches or longer as desired. The advantage of having one of mild steel tube is that sections may be screwed on for lengthening.

The origin of the name is difficult to determine; it could be that the shape of the knob suggests the rubber teat given to babies to suck and sometimes called a dummy.

Bossing Pin

Buchan called the dummy a bossing pin; in bossing lead it is used for bossing against with the round dresser.

EAR

In England and Wales, a projection cast integrally with, or attached to, a pipe socket. It is provided with a hole for fixing with pipe-nail or screw. Cast iron rainwater pipes, soil pipes, etc, may be had with or without ears.

EAVES GUTTER

Eaves are the edges of a roof which overhang the sides of a building—'eavesdrop' means to stand within eaves' drop, to listen secretly. As defined by BS 4118, an eaves gutter is a channel which is fixed at the eaves for collecting rainwater from roofs.

But this BS term should not be taken too literally; eaves gutters are sometimes attached to buildings which are not eaved (see MOULDED GUTTER).

The term probably originated in London, or at least the south east of England, and has crept west as far as Cornwall, to Wales, and north beyond the Midlands.

Back in the days when carpenters made wooden gutters, the plumber was often called in to line them with lead or zinc. Later, plumbers made such gutters entirely of lead or zinc. When cast-iron gutters appeared during the nineteenth century, plumbers despised this unyielding metal and saw that the time would come when their skill in the working of lead and zinc would no longer be needed. But some plumbing skill was needed in the hanging of cast-iron gutters; skill with the line to give the proper fall for carrying rainwater away; skill in cutting this new material quickly and efficiently and, as was often the case, the ability to make suitable brackets or hooks to hold the gutter in place.

Very few plumbing textbooks have ever given such instruction in the hanging of eaves gutters (Buchan did so in great detail) and it was as if plumbers had no desire to do such work; gutter hanging is now any man's job—joiner, bricklayer or handyman.

ELBOW

A knee (qv). It is natural, one supposes, that this English and Welsh word (which is not used in Scotland) should be defined in BS 4118 as 'a pipe fitting for providing a sharp change of direction in a pipeline'.

ELBOW BEND

A fireclay drain fitting illustrated and so named by Howie-Southhook. In plumbing terms, the Scottish knee is called an elbow in England and Wales, but in drain fittings the elbow bend is quite different from the knee bend (Fig 20).

Fig 14 Elbow bend

FALL PIPE

May be heard frequently in England meaning rainwater pipe but, so far, has not been heard in Scotland.

FALL PIPE HEAD

In Yorkshire, North Riding, 'fall pipe head' is used instead of rainwater head (qv).

FAUCET

The enlarged end of a gutter or pipe which receives the spigot of the next section. A word used throughout Scotland and heard from several elderly plumbers in South Cornwall. BS 4118 gives: 'faucet (Scottish) a socket' and in parts of England and Wales the word is looked on as a quaint Scottish one for a socket or an equally quaint American one for a tap (see US FAUCET)

Faucet, an old English word, is one of those taken to America and forgotten in England; the OED gives it as Middle English. Ben Jonson (1573–1637) called a tapster a faucet—one who drew liquor from a cask; a bartender. It was also the peg or spigot for stopping the vent-hole in a cask, and later a tap for drawing liquor from a cask.

In 1923, Grey & Marten Ltd, plumbers' merchants of Southwark Bridge, London, issued a catalogue which used the term 'faucet stop nozzle' but in describing other types of gutter fittings the word 'socket' was used. A catalogue issued by MAC Ltd, Bristol, in 1927 used the word faucet with reference to the inside faucet of a moulded gutter, but on the opposite page, still concerned with moulded gutters, the word socket is used. A catalogue received from Allied Ironfounders Ltd in 1968 has 'faucet stop nozzle' and 'faucet stop end', but 'socket' is used for other gutter fittings; there is no good explanation of this out-of-context use of 'faucet'.

From the French *fausset* meaning falsetto or spigot, the connection between faucet, falsetto and spigot is obscure unless one remembers that faucet and spigot were once synonymous, both meaning a tap, and that plumbing terms are often anatomical in origin. In the throat, the fauces are two pillars which form the margins of a space or cavity descending from the soft palate. A pipe faucet is an enlargement or cavity at the end of the pipe. When a singer uses a falsetto voice, small muscles in the fauces are brought into play to control air movement just as a faucet controls the flow of liquid. In Scottish, 'fause' and 'fauce' mean 'false'.

FERRULE

A pipe fitting for connecting a service pipe to a water main. There are bent ferrules, screwdown ferrules and swivel ferrules, but none of them is a ferrule in the true sense of the word. In Scotland and Cornwall this pipe fitting is a 'service union' (see also VIRREL).

The drive-in ferrule was used until 1870 and simply driven into a hole in the wooden or cast iron water mains after the plumber's joint on the outlet end was made. Although the brass ferrule protected the lead pipe from damage, even it was not a true ferrule as it fitted *into* the lead pipe instead of *over* it. The highly sophisticated type of service pipe connections are still called ferrules. This type of work has passed to 'water engineers' employed by the various Water Boards.

Boiler Screws

Under this heading National Price Lists give overflow ferrules, straight and bent, but overflow ferrules are used for cisterns and not for boilers. 'Boiler screw' is frequently used in the London area and although this is a fairly light brass fitting, it is soldered to lead pipe for connecting to hot water tanks and to cold water tanks. BS 4118 describes a ferrule as a pipe fitting for connecting an overflow pipe to a cistern, but strictly this cannot be correct. (See VIRREL)

Socket Ferrule

The man in the street could be excused for thinking that

this is a ferrule for fitting round the outside of a socket to strengthen it. It is, in fact, a short piece of cast-iron pipe with a screwed brass inspection plate fitted at one end, and this 'ferrule' is caulked *inside* the socket; it is really a spigot with a brass inspection plate. Therefore, 'inspection spigot' would be an exact description, even if rather obvious. British Standards and makers of these goods present them as 'socket ferrules'.

FINGER WIPING

A Scottish phrase sometimes used to indicate what has been described as the cup joint (qv), and in that sense used in a derogatory manner in the past. However, a finger wiping may also be a soldered seam, as in a lead-lined tank, which is an orthodox method of soldering. The term is from the style of wiping the solder to shape with a small, narrow finger cloth of moleskin, the index finger and middle finger pressing the cloth while wiping.

FIRE PLUG

Once upon a time, before standardisation went mad, an 'FP' sign on a house or at the side of the road meant that there was, in case of fire, a valve in the road nearby. This term, of historical interest, has been done to death by H for Hydrant, and modern enamel signs have replaced the old FP plates. The house in which the author lives is scheduled as being of historical and architectural interest; nevertheless, a yellow and black enamelled plate has been fixed to the wall. Out on the road, however, the ancient surface box, now called a hydrant box, has the old FP imprint.

The original fire plugs were wooden plugs driven into holes in the cast iron water mains and, in the event of fire, a plug was knocked out and the water allowed to gush forth to be used by the fire brigade. The old FP sign, like the modern H sign, gave the distance to the plug for easy location. In the mid-nineteenth century there were approximately 30,000 wooden fire plugs in the streets of London.

Even children in Scotland, back in the 1920s, knew the meaning of 'FP'; there was a short rhyme concerning it:

> FP, fire plug
> I'm the lad to pull your lug.

The rhymer would finish by giving another child's ear a vigorous tug.

Fire Point

When the 'FP' imprint on the surface box was pointed out to a member of the Ely Fire Brigade, he said that it meant 'Fire Point', but on having his memory jogged he recalled that it really stood for 'Fire Plug'. An indication as to how old meanings can be forgotten and replaced with new.

FIRE POT

Probably obsolete, but an alternative term for the plumber's devil in the early part of the twentieth century and late nineteenth. Historically, a fire pot was an earthenware pot which, when filled with combustible material, was used as a missile. (See also DEVIL)

FITTER'S UNION

A union used in connecting screwed mild steel pipes. As this term, heard in Sussex, suggests, the union is mostly used by pipe fitters. (See Fig 15)

Barrel Union

Heard in the London area and several parts of the southern half of England, this name indicates that the fitting is used for connecting barrel (qv).

In the union family are a few examples of trade names which are almost part of the plumber's vocabulary.

Dart Union

Known in Britain before the Second World War, this fitting

has been replaced by the Navy Union described below. The only explanation for the name 'Dart' is that it might have been the inventor's name. The Dart Union is still made by Crane in the USA.

MAC Union

Made by Le Bas Tube Co, this has spherical seats which give a sound joint between pipes at any angles within 6° off centre. Inquiries as to the origin of the name have been unsuccessful.

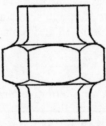

Fig 15   Fitters union

Navy Union

This has two bronze seats, and has been the alternative to the Dart since Crane began producing in Britain all the Crane valves and fittings used in Britain. The name has American roots, being made originally according to US Marine specification—thus, 'Navy'.

Railroad Union

An informant in Essex gave this as a local term for a pipe union. However, the name appears in the Crane catalogue and one can gather from the 'railroad' part of the term that it is of American origin; the railroad in USA is a railway in Britain. In the early days of American railroads Crane produced, with the help of the railroad industry, a union which has been described as 'heavy and ruggedly constructed to withstand strains, expansion, contraction and vibration'. The

Railroad Union has bronze to iron seatings, that is, one seat bronze and the other iron.

FLANK

A Scottish term for a gutter formed at a valley on a roof, and having two sides angled to the slopes of adjoining pitched roofs. Such a gutter may be seen where a slated dormer window or a slated gablet (sometimes called a pedament) meets a main roof at a re-entrant angle.

The illustration shows the section of a lead-covered flank.

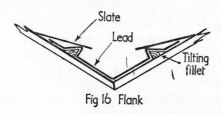

Fig 16 Flank

Valley Gutter

In England, Wales and parts of Scotland, the flank is known as a valley gutter.

FLASHING

In traditional plumbing, a flashing is a strip of copper, lead or zinc, used to weather the junction between a roof covering and another part of the structure. When a plumber flashes or is flashing, he is fixing flashings round a chimney stack, at a dormer window, or at any break in a roof.

In general terms, a flash is a sudden burst of rain or splash of water, according to the OED. The plumber's flashing weathers against such flashes.

FLASHING HOOK

A thumb bat (qv). The 'hook' of the term suggests that

F

something is bent round like a hook, but it is not that kind of hook. It is merely bent over at right angles to the main part, which is the spike. 'Flashing hook' is most likely to be heard in England and Wales.

FLUSHING

Flashing (qv). Like a flash, a flush is a sudden rush of water and that may be the reason for the use of the word 'flushing' by plumbers in rural parts of Kent and Cornwall when speaking of flashings. When questioned about their 'flushings', elderly plumbers are apt to point out that 'flashings' is the correct word. But as a flash is a flush, there is no reason for giving a standard term as the correct one; a standard term can be incorrect, even though British. A look at early methods of flashing rain away from the junction between a roof covering and another part of the structure may suggest that 'flushing' is preferable.

Before plumbers had anything to do with flashings, masons and bricklayers saw to the weathering by laying oversailing courses, sometimes called oversailers. The projecting courses diverted the rain away from the face of the structure on to the roof covering. The junction between covering and structure was sealed with mortar. Instead of oversailing bricks, slates were often inserted in the horizontal joints in chimneys and abutments, the slate in one joint overlapping the slate in the lower joint. They projected about six inches clear of the brick face and tilted slightly to divert water from the face on to the roof covering. A chimney stack with a series of slates on each side, one slate in each horizontal course, gives a pleasant impression of a winged arrangement.

Cornwall, Devon, Wiltshire and Wales (probably other parts of Britain as well) have many old buildings, some going back to the ninth century, with attractive examples of over- sailers, slate and stone. In Cornwall and Devon they are called droppers or dropping stones.

Oversailers are still put into brickwork of chimneys but they are more ornamental than protective, as modern flash- ings provide adequate weathering.

As flashings became the fashion, existing slate and stone oversailers were often cut off *flush* with the face of the structure and flushings inserted in their place. In parts of the West Country, one still may see brick chimneys on terraced houses with slate droppers on one side, while on the other, under different ownership, the droppers have been chopped off and flushed with lead. One can well imagine that plumbers of old would speak of flushing a chimney rather than of flashing it.

FOOT RULE

The rule used by plumbers and others in the building trade has long been known as a foot rule, although it could be two feet or three feet, or even four feet in length, and is not marked in feet.

Blind Man's Rule

A rule, usually made of boxwood, with numerals in thick black type for easy reading.

FOUNTAIN HEAD

A rainwater head in mid-Wales and Herefordshire, but meaning a cast-iron type only. 'Turn a cast-iron head upside down and it looks like a fountain head' was the explanation given.

FROSTCOCK

A type of plugcock (qv) greatly used in Scotland as a main stopcock. The plug component has a small hole at right angles to the main waterway which lines up with a small hole in the body of the cock. Thus, when the cock is turned off, dead water in the service pipe on the outlet side drains through the weep-hole (qv) into the subsoil. By so draining the service pipe, there is no danger of pipes bursting due to dead water freezing.

GARRET WINDOW

A dormer window. Once a watch tower or turret, or a room within the roof of a house, 'garret' is now understood to be an attic. 'Attic window' and (not so common) 'garret window' are more usual in Scotland than 'dormer window'.

GASKIN

An inquisitive trainee may find himself puzzled over this term, but to him it should be an example of how the BS terms and plumbing language do not necessarily make good English sense.

Under one heading the OED says that gaskin is a kind of breech or hose, or it can be the hinder thigh of a horse. Then again : 'Gaskin *rare* = Gasket.' Among other meanings, gasket is given as tow (see HEMP), plaited hemp etc for packing a piston or caulking a joint.

A National Price List gives gaskin as jute, untarred (ie ordinary) or tarred, while under hemp it gives long dressed white jute or Italian silver hemp.

Under gaskin, BS 4118 gives 'spun yarn' and 'yarn' as secondary or non-preferred terms, and describes it as 'loosely-laid plain or tarred rope used for caulking into a spigot and socket joint prior to the insertion of the jointing material'.

But in plumbing circles, gaskin is not quite the same as gasket; the same BS gives 'gasket' as a piece of compressible material, often preformed, used to make a joint between two flat surfaces'. Any mechanic will know that a gasket may be of hard fibrous asbestos or similar material, or even of soft copper. A plumber sometimes makes a gasket of hemp and calls it a grommet (qv).

Davies speaks of driving a gasket of spun yarn into a spigot and socket joint, but he does not use the term 'gaskin'.

'Gaskin' is becoming the fashion in England, but in work-a-day language it is usually 'yarn' and occasionally 'gasketting'. In Scotland it is simply 'rope' used in the right context, or sometimes 'soft rope'. (See also YARNER and also ROPER)

GOOSE BILL

A retired plumber in Billingham, Co Durham, picked up a spoon dresser and said it was a goose bill; like the gentleman with the tongue-stick (qv) he held it out to indicate the similiarity to a goose's bill.

GOOSE NECK

Heard in Cambridgeshire, Kent and Wiltshire for what BS 4118 calls a 'pass-over offset, gutter'. It is a pipe fitting arranged to permit a pipe, like a vent pipe, to pass over an eaves gutter and follow the pitch of the roof. But the goose neck is not necessarily a pipe fitting; the plumber can make a goose neck on pipes of lead, copper, brass, mild steel etc should the necessity arise.

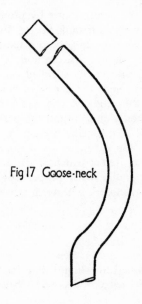

Fig 17 Goose-neck

GRATING

A loose term used in various circumstances and which requires some qualification on each occasion. What is known in England and Wales as a waste fitting—eg, bath waste-fitting or sink waste-fitting—is known in Scotland as a grating; bath grating, sink grating etc. And again, what is known in England and Wales as a wire balloon is, in Scotland, a wire ball grating or wire ball (qv).

As well as the confusion of meanings by qualification, the travelling plumber has to accept that 'grating' may be called a grate or a grid, which are not merely word of mouth variations. In drainage, a surface grating appears in some catalogues as gulley gratings and others as gulley grids. A 1957 Carron catalogue uses all three : grate, grid, and grating, while in a National Price List there are not any gulley gratings, merely gulley grids.

GROMMET, GRUMMET

A ring or wreath of hemp, made by plumbers for use as a washer in pipe fitting and tank fitting. In the making, the hemp is combed with the fingers to dispose of tangle and is then rubbed with tallow. With an end of the hemp in one hand, the rest of the length is rolled on the thigh until a stout string results. Near the centre of the length an initial ring is made to the required size and the string on either side of the ring then wound round the perimeter in opposite directions until all of the string is contained in the grommet.

In many places the second vowel is uttered in such a way that it is difficult to detect which of the alternative spellings is used, and occasionally when 'proper English' is attempted, one might hear 'grimmet'. The term is well known in most parts.

GROUT

A mortar mix thinned to a liquid state for filling in joints on tiles, for filling in urinal stalls, and for making spigot and socket joints in pipes. When used for cement joints in pipes, a piece of gaskin is soaked in the grout before staving into the joint. It is quite usual to wash over cement joints in fireclay pipes with grout after the joint has set. In this thin state grout is sometimes called slurry.

The word is from 'groat', meaning a coarse wholemeal porridge. However, what might be called a grout was commonly used in Glasgow, at least, to seal leaks in lead-lined storage tanks which would have been expensive to replace. The plumber would take along $\frac{1}{4}$lb pea flour, known in Scotland as 'pease meal', mix it to a thin state with water and then pour it into the tank. The particles found the leak and, having sealed it, set very hard. It was necessary to make sure that no water was drawn from the tank until the pease meal had stopped the leak.

GUTTER HEAD

Not the head of a gutter; rather, the head on a rainwater pipe at gutter level—ie a rainwater head—in Westmorland.

HAFFIT

Chambers gives haffet, haffat or haffit to mean the side of the face; the temple. In Scottish plumbing a haffit may be any covering of lead or zinc over what might be compared with a cheek, like a dormer cheek. In certain situations a small flashing has to be fitted at the end of a rone (qv) where the end is at a break in a roof, or at the bottom of a skew (qv), and such a flashing would be a haffit.

HALF-ROUND GUTTER

If any arc more or less than a semicircle can be said to be half-round, then it is a fact that a half-round gutter is, as defined by BS 4118 'an eaves gutter with a half-round cross-section'. The construction lines show that this gutter, which in earlier days was called either a semicircle or half-circle gutter, is less than half round.

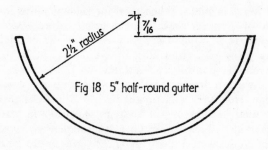

Fig 18 5" half-round gutter

Deep Half-round Gutter

If British Standards allow that a gutter less than half round is half-round it is natural that makers of cast-iron gutters should have in their catalogues a 'deep half-round', not to signify that this type of gutter is a true half-round, but that it is half-round-plus.

Beaded Half-round Gutter

The technical jargon for the bead on this type of gutter is 'integral beading' which means simply that in casting iron gutters a thickening of the metal on one edge, or both, is allowed for. In many parts of England and Wales this type of gutter is not used, the plain (see below) being preferred, probably for cheapness. But from the plumber's point of view, the beaded gutter is more easily cut and that offsets any extra cost.

Place a length of gutter on a solid base and with sharp cold chisel and hammer hit sharply, at the required length, the bead on either side of the down-turned gutter, and then on the centre of the upturned side. The gutter will break neatly across.

Plain Half-round Gutter

Very much used in England and Wales. To cut, one must nick with a hack-saw two opposite edges at the required length and then cut a saw-drift through the thickness of the metal on the upturned bottom; the gutter should snap cleanly when rapped sharply on solid ground. Many plumbers expend much energy and several saw-blades by cutting a deep saw-drift round the outer surface of the gutter before daring to snap it—with such a method the snapping point is one of trial and error.

HATCHET BOLT

A soldering bolt (see COPPER BOLT) with a hatchet-shaped head. When this type of bolt is indicated, the word 'hatchet' replaces 'copper' in any of the variations—hatchet bolt, hatchet bit or hatchet iron. With the hatchet bolt, the 'cutting' edge is tinned on either side to give a V-shaped face. Buchan speaks of the straight bolt as a plumber's bolt, and the hatchet bolt as a gasfitter's tool as, by virtue of the hatchet-shaped head and V-ed face, it is more suitable for soldering joints on light-weight pipes.

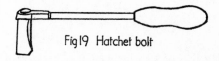

Fig 19 Hatchet bolt

HEAD

An almost nationwide abbreviation of rainwater head. This seems to be the ultimate in abbreviations as far as the rainwater-pipe head is concerned.

BS 4118 does not relate the word 'head' to rainwater heads, but rather as 'a measure of the potential and kinetic energy of water, expressed as a linear distance'.

HEMP

As the plumber knows it, a strong fine fibre used for packing joints; hemp is used to make grommets, washers, and, on screwed pipework for water, a few strands wound into the male thread ensures a water-tight joint.

Italian silver hemp is best for the plumber's purpose, but jute, a cheaper and less durable material, is often passed off as hemp. The 'jute' under the heading 'Hemp' in price lists is easily recognisable by its coarseness compared with the fine hemp. Just as jute is often used as a substitute for hemp, good quality hemp is used, in other industries, as a substitute for flax.

Hair

A work-a-day term in various places; a teased out hank of hemp is like a hank of hair. Thus the plumber's quip when preparing a male thread : 'It goes in better with hair round it.'

Tow

Another work-a-day term for hemp and more likely to be heard from elderly plumbers. But the word is not mere slang; somehow it has found its way from the factory floor where the processing of hemp stalks takes place; it is cleaned in a 'tow machine'.

HIP

'The meeting line of two inclined roof surfaces which meet at a salient angle'; so says BS 2717.

Hips may be weathered with ridge tiles or, as in the case of mitred hips, with soakers; the plumber has no work to do in either of these methods. However, where neither tile nor soaker is used, he may be called upon to weather the hip (and the ridge) with zinc ridge or ridging, which comes ready shaped and in standard lengths, or, less likely nowadays, he may have to do the weathering with sheet lead. (See also PIEND)

HOLDFAST

Scottish, meaning pipe hook (qv). While the English plumber is more used to eared cast-iron rainwater pipes, waste pipes, etc, which he fixes with pipe-nails, Scottish plumbers and to some extent Welsh plumbers use such pipes without ears and hold them fast with pipe hooks or holdfasts as the case may be. The practice of fixing pipes on outside walls with holdfasts probably stems from the days when the majority of buildings were of stone, and it was easier to move up or down to a suitable joint in the masonry into which to drive the holdfast; there is no come or go, up or down, when using eared pipes.

In South Wales, 'holdfast' was given to mean a soft block, of breeze or wood, built in with the brickwork to give a fixing place for pipe brackets etc.

HOPPER HEAD

BS 4118 describes this very well: 'A flat or angle-backed rainwater head of tapered shape.' The term is well known from the North of England to the South, and in parts of Wales. However, the sense of the descriptive word is lost in places and the term is applied to rainwater heads of any shape. The modern plastic head of rectangular cross-section is often called a hopper and in East Anglia 'square hopper' has been heard. (See Fig 29)

INCISION JOINT

When one lead pipe branches into another and a solder joint is made at the branch, the joint is called a branch joint or an incision joint: 'incision' is less popular than 'branch'. In preparing the joint, the pipe into which the other branch is pierced or slit as the case may be and the hole tafted (see TAFT) out to a suitable size and shape to receive the other. 'Incise' is to cut into; thus the term 'incision joint'. Although the term is used only with reference to this soldered joint, it could well refer to the modern method of cutting into plastic soil pipe and waste pipe to make branch connections.

INITIATION

A custom which probably died out shortly after the Second World War was the initiation of apprentices within a few weeks of the start of their term of training. Once he had been initiated, or 'smudged' as it was usually called, the apprentice was fairly safe from similar attacks by his workmates. The ceremony consisted of the downing of the apprentice and, once his workmates had him helpless on the floor, his dungarees and any underclothes were pulled down and his navel and penis liberally coated with smudge (qv). Very old plumbers tell of the days when the initiation included the singeing of pubic hairs with a hot soldering bolt.

INTERCEPTOR

Also known as an intercepting trap and described by BS 4118 as 'a trap fixed on a drain to prevent the passage of sewer gas into the drain'. The English style of trap is usually incorporated in a manhole, so as well as intercepting the sewer gas it serves as a convenient place to intercept any cause of blockage.

During the last twenty years or so, interceptors have been somewhat miscalled because of a tendency to block up, and instead of looking for the cause of the complaint the majority of local authorities have decided to eliminate interceptors by taking untrapped drains into sewers. The real cause of faulty traps could be the increasing army of do-it-yourself plumbers, etc, who take up manhole covers and, in general, interfere with house drainage.

JAMNUT

In Scotland and the North of England near the Anglo-Scottish border, the backnut (qv) is better known as a jamnut; it jams against the fixture, say, to secure a tap or fitting, and in the case of a pipe connector (see CONNECTOR) it jams against the grommet and socket.

Also heard in the West of Scotland was 'checknut', but the backnut, as used by plumbers, is more of a jamnut than a checknut which, in engineering, has more of the locknut meaning only.

JAW-BOX

A cast-iron kitchen sink very common in Scotland from about the 1870s until the 1940s, when the affluence of war years brought about great clearing out of jaw-boxes and replacement with white fireclay sinks; there may be a few jaw-boxes still in use. Buchan records that they were in general use in houses of the lower and middle classes of Glasgow. In the vernacular the term continued to be used with reference to fireclay sinks.

Davies says quite definitely, but wrongly, that the jaw-box was so named on account of the jawing of washerwomen while at work. The term is self-explanatory when one remembers that the early sinks of stone and later of fireclay were called splash-stones and that a jaw or jawp, in Scottish dialect, is a splash or dash of water—Scottish ladies frequently lament jawping their stockings in wet weather—thus the jaw-box was a cast-iron box for containing jaws or jawps.

JUMPER

(a) In a screwdown valve the spindled disc to which a fibre, leather or rubber washer is fixed. Called a jumper probably because it jumps up off the valve seating when the valve is opened, due to the pressure of the incoming water. On certain screwdown valves the jumper is riveted to the worm, and in this case it is known as a 'fixed jumper'; it does not jump but is raised and lowered by the action of the worm, independent of water pressure.

(b) A long cold chisel for cutting away brickwork or masonry. Also, a type of chisel, often with what is known as a starred cutting end, for cutting holes in brickwork and masonry. The jumper is turned constantly whilst being struck with hammer and the starred cutting edge makes the hole as would a drill. Small plumbing concerns to whom expensive brick drilling equipment would be uneconomical because of infrequent use, can improvise quite satisfactorily by cutting, with a hacksaw, teeth on the end of a piece of mild steel tube of ½in to 1½in bore; when used as a jumper the serrated end will make a neat round hole in brickwork or in masonry.

KNEE

The Scottish name for a pipe fitting in the shape of a short radius bend with threads at each end, either male or female. They are used usually in conjunction with screwed mild steel tubes.

Most plumbers with many years of plumbing behind them are familiar with the rhyme :

> He's a handy man, the plumber,
> in winter and in summer.
> He can make a joint with ease
> and handle female knees;
> that *is* his line—he's the plumber.

In Scottish terms a female knee is a fitting with a female thread at both ends. However, in England and Wales, where the above rhyme is as well known as in Scotland, the word 'knee' is not known in the Scottish sense and thus the rhyme contains in the fourth line a mere silly reference to the knees of the so-called gentler sex. The knee is known to English plumbers as an elbow (qv)—there is no fun in handling a female elbow.

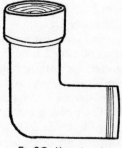

Fig 20 Knee-bend

KNEE BEND

With reference to fireclay drain fittings, the knuckle bend (qv) is often referred to as a knee bend. A catalogue by Howie-Southhook differentiates between the knee bend and the knuckle bend. (See also ELBOW BEND)

KNEE ROOF

A mansard roof. The term used by an elderly plumber in Suffolk; the rafters of a mansard roof are bent (see also CURB ROOF) downward at the lower ends, thus giving the knee outline, and they are known as knee-rafters.

KNOB IRON

A plumber's iron (qv) has been referred to as a knob iron in Cornwall. The knob, the soldering end, obviously suggests the name.

KNUCKLE BEND

Throughout Britain, short radius bends for fireclay drain-work are often referred to as knuckles or knuckle bends, but although a knuckle is indeed a short radius bend, a short radius bend is not always a knuckle. Howie-Southhook differentiates between the two. (See also KNEE BEND)

Fig 21 Knuckle-bend   Fig 22 Short ¼ bend

KNURL

The milling, or notching, round certain nuts or rings, like packing glands (or round certain silver coins) is said to be knurling and so a knurl is a very small notch. It follows that in plumber-talk a knurl, particularly in Scotland, became slang for a vague, fractional measurement—almost infinitesimal. A pipe might have to be moved or turned 'just a knurl' to be in correct alignment, say.

The OED says that knurl or nurl may be the diminutive of knur or knurr which means, amongst other things, a wooden ball used in the north country game of 'Knur and spell'. One is never quite certain if the expression 'north country' is meant to include any part of Scotland, but it is fairly certain that the knur referred to in the OED has little connection with another slang term meaning knurl—ball hair (qv).

LANDERS

Name for 'eaves gutter' in North Wales. In the smelting industry, when a furnace is tapped, the stream of molten metal pours out and *lands* on a trough or gutter called a lander which conveys the metal to the next stage of the process. Thus, lander, from local main industry, is applied to eaves gutters in plumbing terms, but should not be confused with launder.

LAUNDER

Name for an 'eaves gutter' in Cornwall. In the days when free tin could be panned from the soil, wooden troughs or shutes were made to channel water for washing or laundering of the ore. The troughs, called launders, were similar to the gutters attached to the eaves of buildings and these were fixed by carpenters, followed by plumbers who lined the launders with lead or zinc. The name 'launders' has stuck from those early times, through the lead and cast-iron plumbing ages to the present plastic age; nowadays there is plastic launder or laundering.

An informant in Bath gave 'launder' as a Bristol-based term, but as there has been much corruption in dialects, landers (qv) and launders are frequently confused as being corruptions one of the other. While conducting a tour of iron works in Bedford, the guide pointed out the lander as a 'launder'. Information from the USA gives 'launder' as the form used in American steelworks.

LEAD

For so long have the properties of lead (see also PLUMB) been used for padding technical books that they have become a kind of senseless jingle which no longer bears repeating.

Lead was probably discovered in most distant times by primitive man who happened to notice molten metal trickle from his fire. The alchemists gave it the name Saturn and used the Saturn symbol.

Pliny (about AD 23–79) treated lead and tin as *plumbum nigrum* and *plumbum album*, two varieties of the same species. Red lead and white lead as we know them today were also known in the eighth century.

If we think, as we look at the Roman plumbing in the city of Bath, that the Romans were ahead of the times, it is likely that we have forgotten about Moses who, as we are told in Exodus, used many non-ferrous metals including lead in the making of the Tabernacle, about 1491 BC.

Leed

In Scottish vernacular, lead is pronounced 'leed'; leed pipe, leed sheet, etc. The graphite in a pencil is likewise leed pencil. The Lead Hills in Lanarkshire, where the Romans mined lead, are Leed Hulls—hills become hulls.

LEAD HAT

In work-a-day terms, a usual name for a lead slate (qv) when made for flashing round a cast-iron pipe. Before it is

G

fitted to the pipe, particularly by the Scottish method, the lead slate has the appearance of a hat or a fireman's helmet.

LEAD SLATE

Sometimes referred to simply as a slate piece. When a pipe passes through a pitched roof, the necessary number of slates

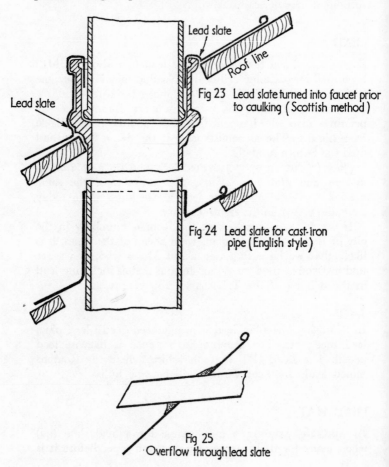

Fig 23   Lead slate turned into faucet prior to caulking (Scottish method)

Fig 24   Lead slate for cast-iron pipe (English style)

Fig 25
Overflow through lead slate

or tiles are removed or omitted and a piece of lead fitted in their place. The lead is previously worked to a shape which, when the pipe passes through, acts as a flashing to the inclined intersection between pipe and slates. Lead overflow pipes and expansion pipes (see SWAN-NECK) may pass through a lead slate and the intersection sealed with a solder joint or by lead-burning.

The most common method of fitting lead slates in England and Wales is inferior to the Scottish way. In the south, the weather-proofing of the flashing depends greatly on a packing of putty, bitumen or similar material. The Scottish method requires more skill and results in the lead slate being what might be called an integral part of the pipe. The illustrations also show a lead overflow passing through a lead slate.

LIGHTNING FLASHING

In Billingham, Co Durham, a retired plumber referred to herringbone flashing as lightning flashing; with a little imagination one can see that the herringbone flashing could be similar to a cartoonist's impression of a lightning flash.

LONDON FLASHING

When asked about herringbone flashing, a retired plumber in the North Riding called it London flashing and declared that it was not the right way to do steps. (See STEP FLASHING)

LONGSCREW

BS 4118 gives this as a piece of mild steel tube threaded externally at each end, one end having the thread sufficiently long to accommodate a backnut and the full length of a socket. It is used to join together two pieces of steel tube, neither of which can be rotated.

It is true to say that a longscrew, more often than not

called a long thread, may have a thread long enough to
accommodate a backnut and a socket, but it is equally true
that plumbers make longthreads to meet requirements from
job to job; eg to accommodate two backnuts, a grommet,
and a thickness of metal, when connecting mild steel tube to
a storage tank; also, ballvalves, pillar taps, etc, are made with
longthreads or long threads, longscrews or long screws.

The terms 'longscrew' and 'longthread' (see also RUN-
NER) are frequently used as synonyms of connector (qv),
but a longscrew does not become a connector until it is furnished with a backnut and a socket so that it can be used
to connect together two pieces of mild steel tube. The
National Price List shows longscrews separate from connectors.

LUKIM

A dormer (qv) window in old Suffolk dialect. Probably a
variation of luthern for a dormer window, the French
lucarne or Gothic *lukarn*.

LURK

A Scottish word meaning a crease, wrinkle, fold or crevice.
An English or a Welsh plumber might make a wrinkle, kink
or crease in a piece of sheet lead or a piece of lead pipe,
whilst working them—a thing to be avoided. The Scottish
plumber would say he had made a lurk or that he had
lurked the lead.

LUTING

A word heard only once with reference to an eaves gutter
in a small East Anglian village, but there has not been any
verification of such a word with this meaning. It is feasible
that the word should have been 'fluting' which, in architectural terms, is a furrow or channel in stone, and any word
meaning channel or groove could well be applied to eaves
gutters.

MANDREL

Sometimes spelt mandril and (incorrectly) mandrill; the latter, according to the OED, is 'the largest, most hideous, and most ferocious of the baboons . . .' Nevertheless, the word was spelt this way in a tool list issued in 1966 by the Associated Master Plumbers.

Mandrels are cylindrical lengths of boxwood, beech, or other hard wood with diameters all corresponding to the standard bores of lead pipes, soil and waste.

Bending Mandrel

Better known as the long mandrel, this is about one yard long (or, say, 1,000 mm) and it is called a bending mandrel because it is frequently used as a lever when bending lead pipes. It is also useful for driving through short lengths of lead pipe to remove dents and kinks. But apart from this purpose, the long mandrel was once a necessary tool in the making of soil and waste pipes from sheet lead. A long mandrel of desired diameter was laid on the lead strip parallel to the lengthwise edges which were then lifted and the lead bent round the mandrel with the edges meeting. The mandrel was then withdrawn, the edges prepared for soldering and the straight seam was then soldered with copper bolt and fine solder.

Driving Mandrel

Or the short mandrel about six inches long and, having a tapered end, mostly used for driving through lead pipes to remove dents, etc.

Drift

Heard in Cockermouth, Cumberland, with reference to a mandrel. Davies also uses the word 'drift' (qv) as an alternative to mandrel.

Pulling-up Stick

This expression, heard in Pontypool, indicates the use to which the mandrel is put when bending lead pipe.

Toggle

This is not a common word in plumbing but in Billingham, Co Durham, it was used as an alternative to mandrel.

MANSARD ROOF

BS 2717 : 'A roof with two pitches on each side of the ridge, the steeper commencing at the eaves and intersecting with a flatter pitch finishing at the ridge. The term is sometimes applied to a roof with steeply pitched slopes surmounted by a flat.'

This type of roof is named after the French architect who introduced it, François Mansart (1598–1666) and is of interest to plumbers since the intersection of the two pitches may have to be flashed with lead, zinc, or copper. In the case of the type of mansard which consists of pitches surmounted by a flat, the intersection of pitch and flat is finished with a lead-covered torus roll.

METAL

To the layman and to Scottish plumbers this can mean any of the obvious metals—gold, silver, copper, iron, lead, etc, and even alloys—and in some trades metal means, specifically, cast iron. To English and Welsh plumbers 'metal' is specifically plumbers' solder, sometimes called plumbers' metal. But Scottish plumbers may not know what is meant by plumbers' solder or plumbers' metal. It is the solder used to make wiped joints and it comes in bars weighing about one pound. Tinman's solder (qv) or fine solder is *not* metal.

MONKEY DUNG

An asbestos composition in the form of a dry fibrous powder which, when mixed with water to the consistency of mortar, is used as an insulating covering for boilers and hot pipes. The first coat should be applied in small patches until the whole surface of boiler or pipe is covered : that stage is called 'spotting'. Subsequent coats may be applied with a trowel and the final coat given a smooth trowel-finish.

When in the mortar state the material looks like some kind of dung but whether it is similar to monkey's dung is open to question. The term was used quite a lot in Scotland some years ago but in England it has been heard only occasionally. 'Asbestos cement' or simply 'lagging' are more common.

MOULDED GUTTER

This BS term is ambiguous, as all gutters are moulded, half-round or otherwise, whether they are made by hand, cast in a moulder's sand bed, or pressed in plastic by a machine. But plumbers know that 'moulded' is used in the architectural sense, and that it is an eaves gutter with flat sole, upright back and ornamental front. Moulded gutters may be screwed to the facia board on the eaves of a building and supported by brackets, and in such instances they may be rightly called 'eaves gutters'. However, they are better suited for use on an uneaved building with the flat sole resting on the wall head for support and the ornamental front, the only part to be seen from ground level, presenting a pleasing cornice effect which is enhanced when painted with a colour suited to the natural materials used for the house. 'Cornice gutter' would be a more descriptive name as the cornice effect applies whether the building is eaved or uneaved. (See ORNAMENTAL GUTTER)

NAVY GRAVY

A knurl (qv) in the slang sense. Heard from a retired plumber in St Ives, Cornwall, with reference to a fine measurement. Perhaps sea-going experience explains the implications of this term.

NOSE COCK

A bib tap (qv). Heard in Lanarkshire, Scotland. One may also hear 'bib cran' and 'nose cran' in Scotland.

OFFSET

A pipe fitting for connecting two pipes whose axes are parallel but not in line, or a double bend formed in a piperun to continue it on a parallel course, but not in line with its former course.

Fig 26 Offset

Set-off

Many plumbers, middle-aged and older in particular, use this term for the offset, and as the pipe is indeed set off its course, 'set-off' is more sensible than 'offset'.

Swan-neck

Used in England and Wales for 'offset', more so in the southern half of England (cf GOOSE NECK).

OG GUTTER

Sometimes written 'ogee gutter' which, being an English term also adopted by Wales, has subsequently become a British Standard term. Almost since the inception of cast-iron gutters the OG gutter has been defined in terms which have become standardised to mean an eaves gutter with a combined sole and front of ogee shape.

In this age, when the argument in favour of standardising

plumbing terms is that plumbing is a highly technical and precise subject, one would think that instead of perpetuating a misnomer, any modern definition would point out that the OG gutter does *not* have an ogee cross-section. One defensive retort by the progressives is that any double curve is, in work-a-day terms, an ogee. Not so.

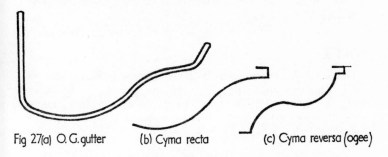

Fig 27(a) O. G. gutter   (b) Cyma recta   (c) Cyma reversa (ogee)

An ogee outline is one consisting of a double curve, convex above and concave below, and, in architectural terms, known as *cyma reversa*. The opposite, which is concave above and convex below, is *cyma recta*. Comparison with a cross section of an OG gutter suggests that, if anything, the gutter has the *cyma recta* shape and not the *cyma reversa* (or ogee) outline.

Notts OG gutter

Makers of cast-iron gutters have this type of gutter in their catalogues and, as the name suggests, it is probably used more in the Nottingham area than in other parts. This varies from the standard OG gutter in that it has a flat sole and upright back and is deeper, but the misnomer regarding the shape of the front is perpetuated. The Notts OG gutter answers to the description of a moulded gutter (qv).

Scotch OG

Heard from an elderly gentleman plumber in Loughborough, near Leicester, with reference to the standard OG gutter in distinguishing between it and the Notts OG gutter, which is used to some extent in the Leicester area. But why he should refer to it as a Scotch OG is rather strange as the English standard OG gutter is little used in Scotland. However, he was correct with the term, but had he said : 'Scotch OG gutter' he would have been incorrect. He was not aware that the Notts OG gutter is also a Scotch OG.

For further on the intricacies of this term see ORNA-MENTAL GUTTER.

ORIEL

The oriel is seldom seen in modern house-building and so the English or Welsh plumber who travels to Scotland may be mystified by the term. In the Scottish building trade 'oriel' means a bay window (qv) whether it is on the ground floor or on an upper floor.

ORNAMENTAL GUTTER

While Davies was writing in London about OG gutters, Buchan, in Glasgow, referring to the same gutter as Davies, was calling it an ornamental gutter. To Scottish plumbers the letters OG, as well as meaning ogee, stand for 'ornamental gutter' and an OG gutter (qv) would be an ornamental gutter with a true ogee outline. But 'OG gutter' when applied to any other shape would be nonsense—it would mean 'ornamental-gutter gutter'. So it follows that the so-called moulded gutter (qv) is just another OG to the Scot, although the term 'moulded gutter' is frequently used.

PACKING IRON

A stemmer (qv). The stemmer is used for packing yarn into a spigot and socket joint, so the tool is also called a packing iron; a term heard in Hawick.

PANE

The balled or edged end of the head of a hammer. The

cross-pane is an edged end at right angles to the shaft and the straight pane an edged end in line with the shaft: the term 'ball-pane' is self-explanatory. Traditionally, the plumber's hammer has a straight pane, but many plumbers have one of each type and of different weights. Ball-paned hammers are often referred to as engineers' hammers, but engineers probably use all three types.

In Scottish dialect 'pane' becomes 'peen'. For example, a peen of glass for a window, and ball-peen hammer, straight peen, etc. 'Peend' or 'peened' is often wrongly used for the peen of a hammer. (See also PIEND)

PAP

In Scottish plumbing, a pap is the nozzle outlet of a gutter— eg centre pap is a centre outlet. Buchan, too, referred to such a gutter nozzle as a pap. The outlet of a closet pan can be a pap, as can the bulge on the underside of a sink or basin outlet. A common pronunciation is paup or pop—nothing to do with pop-up-wastes.

The word is supposed to echo the sound of an infant feeding and means a teat or a nipple. Scottish vernacular extends the meaning to the female breast, and so a well-shaped female may have a nice pair of paps.

PEERIE

The turnpin (qv) is similar in appearance to a child's cone-shaped spinning top; such a top is known in Scottish dialect as a peerie. Thus, the name is given to the turnpin occasionally in work-a-day language.

PELTIE

Heard in Coupar Angus, Scotland, for a hammer.

PIEND

BS 2717 gives this as the Scottish name for the hip (qv) of

a roof, but the Scottish plumber or slater usually refers to the hip as the ridge or 'rigging'.

Piend is pronounced 'peen' or 'peend'. Under 'peen' the Scottish National Dictionary gives alternative spellings: pien, pean, pin(e), pi(e)nd, peind, peand and pen(d). In Scottish dialect a 'peen' can be a pin, a pane of glass, or the pane of a hammer (see also PANE). But the French *pente*, from Latin *pendere* meaning pitch (of roofs), side, wall (of tents) suggests that 'piend' could be from the French and quite independent of pin or pane. Scottish carpenters talk about 'piend rafters', ie the rafters in the triangular sloping end of a hipped or piend roof, and the hip of a roof is referred to as the piend-ridge. The piend, then, is not the hip, but what BS 2717 describes as a hipped end: 'a roof surface, usually triangular, bounded by the hips at the sides and the eaves at the base.' Like 'hipped roof', 'piend roof' differentiates between this type of roof and a gabled roof.

PIG'S LUG

The Scottish term for an English dog ear (qv). The making of such a corner in sheet metal work is called pig-lugging.

Sow's Lug

Also in Scotland, when an external corner is bossed in sheet lead the surplus lead at the top of the corner is called a pig's lug and, frequently, sow's lug.

PILLARED COCK

Buchan used this term when describing his method of fitting cocks for bath and basin supplies. The cocks, hot and cold, were connected to one outlet pipe. They were situated near floor level and the *heads* were pillared, in that the rods were fixed to the heads and taken up through the wooden or marble bath or basin top. Ornamental plates or escutcheons covered the holes in the surface material and handles, hot and cold, fitted on to the top of the pillars.

PILLAR TAP

A tap, suitable for mounting on a horizontal surface, having a vertical inlet and a nozzle bent to discharge in a downward direction—this definition is from BS 4118. This 'screw-down tap' (see TAP) has been used mainly for baths and lavatory basins in the past, but in recent years they have also been fitted to the popular metal or fibreglass sink unit.

One is tempted to think that the 'pillar' is the vertical inlet or long threaded stem necessary on such fittings, but the word is probably a perpetuation of the 'pillared cock' of the nineteenth century.

To the Scottish plumber, the pillar tap is a cran (qv), bath or basin, as it has a nozzle bent in a downward direction.

PIN-HOLE

This is what the name suggests—a pin-hole made in a lead water pipe preferably with a panel pin or similar small nail. A solder joint cannot be made if there is water in the pipe and there are times when a main stopcock does not quite shut off, or when there is water in a dead leg of pipe. One method of getting rid of this water is by pin-holing. Once the water has piddled to a stop the joint can be made. Pin-holing is not taught at technical colleges because learned plumbers consider it bad practice, yet there is nothing wrong with a properly sealed pin-hole.

One wrong way of sealing the hole is to drive in a small wooden peg, cut it off flush with the pipe and leave it at that; if it does not weep immediately it will do so very soon. Another wrong method is to carry the first method a stage further and taft the lead over the peg with a blunt chisel or the straight pane of a small hammer; such a job may last for a very long time—until it splits open when there is movement in the pipe.

A pin-hole can best be sealed by driving in a small and short hardwood peg as before, but then, clean the lead round the peg and solder it over with fine solder and soldering bolt;

the soldering with such localised heat is possible, whereas the wider heat of a blowlamp generates steam which blows before the soldering can be done.

Discovered in London in the early 1960s was a truly original example of pin-holing. A handyman-plumber had trouble shutting the main off, and to relieve the pipe on which he was working of water he made several sizeable holes. Having made his solder joint (sic) he then screwed ¾in x no 8 black japanned woodscrews into the pin-holes— five in number. They rusted away within a few months.

Pricker-hole

Heard in the North Riding for 'pin-hole'. (See also WEEP-HOLE)

PIPE HOOK

In England and Wales mainly, a hooked spike with the hook designed to grip the pipe when the spike is driven into a wall. Pipe hooks are made for all sizes of pipe, lead and iron, from ¼in to 4in in diameter. (See also HOLDFAST)

PLUG

The phrase 'pull the plug', as used to denote the flushing of a high-level closet cistern, is a relic of the days when closets were flushed by pulling or pushing a handle which in turn pulled a wire, the other end of which was attached to the plug of a drop-valve in a storage cistern. The plug was usually of brass and had a ground seating or a leather washer. Once the flush was started, the water flowed into the pan until the handle was returned to its original position and the plug allowed to drop back into the seating at the top end of the flushpipe.

Tampin

In Roxburghshire, Peebleshire and parts of Northumberland, the tampin is called a plug (see also DOOK).

PLUG COCK

A taper-seated cock (see VALVE) in which the plug is held in position by means of a washer and nut on a screw on the bottom of the plug. This type of main stopcock is used in Scotland and parts of England, but in North Devon the cock which is referred to by Water Board workers as a plug cock is in fact a gland cock.

Gland Cock

A cock in which the plug is held in the body by a packing gland.

PLUGGING CHISEL

An English and Welsh term for a thin-bladed cold chisel with splayed blade for raking out or cutting out mortar joints in brickwork for the purpose of receiving wooden wall plugs. In this respect the plugging chisel serves the same purpose as the Scottish dooking iron but the plugging chisel is useless for cutting holes in masonry or brickwork. The plugging chisel is more likely to be called a raggling iron (qv) in Scotland. When the average English plumber cuts a hole in brickwork or masonry he would probably use a cold chisel or a starred chisel or jumper (qv). If a cold chisel is used the hole must be cut a little wider than the blade of the chisel to avoid binding.

PLUMB

Diminutive of 'plumber', used by other building workers, like 'brickie' for bricklayer.

It is appropriate that in mid Wales, where lead (qv) was mined for many centuries, plumbers still speak of 'plumb', from Latin *plumbum,* meaning lead. The old conundrum : 'If a brickie lays bricks why can't a plumber lay plums' is rather off the mark because plumbers lay plumb when laying lead gutters on roofs in North Wales.

In everyday terms, plumb has little of the lead meaning;

it means vertical—straight down—but seldom straight up. As a verb it is to make vertical; but whether a pipe is made vertical—ie plumbed—by means of a plumb-line (see PLUMB BOB) or by casting the eye up or down the stack does not matter—it is still plumbed. To save himself the trouble of going down a ladder, a plumber who is fixing outside pipes will shout to another : 'Plumb this, mate, while I fix it' or he may say : 'Cast your eye up this, mate.'

To plumb a house is to install a system of pipes, whether the pipes are of cast iron, mild steel, lead, copper, or plastic. The plumbing of a house does not seem to include laying of gutters, flashing to chimneys or dormers, etc. The average layman knows little of the amount of plumbing on a roof.

The literary expression 'plumbing the depths of despair' is probably borrowed from the old nautical method of fathoming the sea by heaving overboard a line with a hunk of lead tied on as a sinker. The only time the plumber is likely to use the idea of 'plumbing the depths' is in the rude limerick, well known in the trade :

> There was a young plumber of Dee
> Who was plumbing his girl by the sea.
> 'Oh, plumber, stop plumbing,
> There's somebody coming !'
> 'You're right,' said he, 'It's me.'

PLUMB BOB

Plumb is lead (qv) and 'bob' according to the OED is, amongst other things a knob; a plumb bob is a lead knob. For some years plumb bobs have been manufactured, but they are usually of brass or iron—one is unlikely to hear of a brass bob or an iron bob, they are always plumb bobs. When a bob is attached to a chalk line the line is then a plumb-line as well as a chalk line.

When scrap lead is available, plumbers can easily make plumb bobs. Press a turnpin or tampin (see TAMPION) point first into dampish sand, making sure that the top of the turnpin is level. Withdraw the turnpin and down through the centre point of the mould press a bent nail or stout wire. Pour molten lead very slowly and carefully into the mould—gently, because of the danger of a blow-out due to dampness in the sand mould. The result is a conical bob with hook, on to which a line can be tied.

PLUMBER

The OED describes the plumber: 'An artisan who works in lead, zinc and tin, fitting in, soldering, and repairing water and gas pipes, cisterns, boilers, and the like in buildings; orig. a man who dealt and worked in lead.' It is interesting that the OED, like the layman, associates the plumber mainly with pipes, cisterns, etc, and does not mention the more elaborate work on the covering of domes, turrets, spires and dormers with sheet lead, sheet copper, and sheet zinc. Old French *plummier, plommier* (modern French *plombier*) Latin *plumbarius* from *plumbum* meaning lead, but the modern plumber is not so much a *plumbarius* as a domestic engineer knowing little of the old art of working lead—good lead workers are dying off rapidly. In parts of Cornwall and Wales plumbers have given over the flashing of chimneys to bricklayers and masons, and the same pattern shows throughout England and Wales; little sections of the plumber's work being given over to other trades or to semi-skilled—domestic engineering with modern materials is cleaner and more profitable.

Sanitary Engineer

A high fallutin' name which does not signify any plumbing qualifications. Many builders and handymen use the name to describe themselves, and there is no legislation to prevent the use of it by the unqualified. For over a hundred years plumbers have protested about the use of this title—they still protest.

Water Engineer

A name still to be seen over many old workshops throughout England and Wales. It signifies that the firm or the

H

individual specialises in well work, pumps, and the taking of piped water from natural sources.

Bosser

An old nickname for a plumber, meaning one who bosses lead.

Copper Bottom

Davies gives this nickname, insinuating that it is applied to plumbers, but it is more likely to have applied to copper-smiths.

Three Branch Hand

A term which might be used in parts of England, but not verified. Davies uses it to describe a man who was capable of doing plumbing, painting and glazing. He also remarks that though these three branch hands were despised by London plumbers, he thought that some of them could hold their own against good plumbers.

PLUMBER'S CANDLE

Until the beginning of this century the plumber's tallow candle served a dual purpose; it gave him a glimmer of light when required, and he also used it as a flux (see also TOUCH). Inquiries regarding the plumber's candle have been directed at a great number of plumbers, but not one had heard of such a thing.

In course of research, the 'touch' trail led to Price's Patent Candle Company Ltd of Battersea, London, and although the informant had never heard of 'touch' or Russian Tallow (qv) he did disclose that his firm does in fact produce plumbers' candles for use as fluxes. Messrs Price kindly supplied two sample candles. There are two types:

Candle (with tallow)

This contains a proportion of tallow and is made in candle form but without a wick. It measures about eight and three-quarter inches long and about three-quarters of an inch diameter. Several tons per year are supplied to the Post Office Engineering department.

Candle (without tallow)

This candle consists of a mixture of fatty acids and has a wick. It is slightly thicker than the first type and measures from end to tip about eight and one-eighth inches. It appears to burn with a clearer flame than the ordinary wax candle.

PLUMBER'S IRON

Many plumbers who, in recent years, have worked in old workshops may recognise this as an old tool which they, not

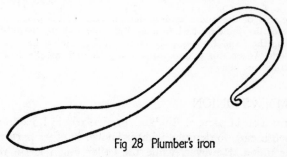

Fig 28 Plumber's iron

knowing its original purpose, have used as a dummy at one time or another. It is really an instrument for soldering as done in the old days—probably up to the beginning of the twentieth century. It was sometimes called a soldering iron and there was, therefore, some confusion in everyday language between 'plumber's iron' and the copper bolt, which even today is called 'a soldering iron' in places.

Of malleable iron throughout, a plumber's irons varied in weight from about one and a half pounds to five or six pounds; the heavier types were often referred to as heating irons because they were used for pre-heating large-diameter lead pipes when jointing—eg lead barrel for well work.

In the making of soldered seams in lead-lined tanks and in lead gutters, the hot solder from the ladle was splashed along the seam, then the red-hot iron, having been scaled

with a file, was drawn along to melt the solder and keep it
in molten state while it was pressed into shape and wiped
with moleskin cloth to form a finished seam.

For the making of an underhand joint, who better than
Hellyer to explain the art of doing so with plumber's iron
or, as he called it at one time, the rosy iron; it is impossible
to tell from the text if 'rosy iron' was a term or if 'rosy'
was just an example of the poetical language he was prone
to use.

If you prefer to use an iron, see that it is well cleaned, and just
red-hot before using; then, having a nice body of solder on the
joint which is to be, rub the iron round and round it, and take
it all off on the cloth. Rub the hot iron upon the solder, and work
it up into a nice consistency upon the cloth, forming it all the while
into the shape somewhat like a sausage-roll, then quickly place the
centre part of the solder against the underside of the pipes, at the
centre of the jointing, and turn up the outer half on the off-side,
bringing the hand back quickly to turn up the other half on the
near side. When this is done, wipe the joint with all the dexterity
you can command.

PLUMBER'S UNION

A term used in parts of England, particularly in London and
the south east, to describe what is known in other parts as a
boiler union. However, while the boiler union has a male
thread on one end, the plumber's union may have a male
thread or a female; the type with female thread is seldom
used. In Scotland, the male plumber's union is better known
as a boiler coupling.

Grey & Marten's 1923 catalogue does not use the term
'plumber's union'; it describes it as 'brass lead to iron union';
the end for iron being either male or female.

MAC's 1927 catalogue shows a lead to lead union, and
a lead to female iron union, both described as plumber's
unions, but the male iron to lead union is described as a
boiler union.

PODGER

A podge is someone, or something, short and fat. A podger

is a short, stout bar of iron or steel used as a lever, particularly for turning certain tools like box spanners, tank cutters and for tightening the jaws of vices.

The term has been heard in Lanarkshire and is probably well known in other parts of Scotland. In England and Wales it might be called a Tommy (qv) or Tommy Bar.

POT

A plumber's pot (sometimes called 'melting pot' is a cast-iron affair, sizes varying from about five to twelve inches in diameter, with three short legs or feet and a handle of wrought iron.

Lead Pot

A melting pot for melting lead.

Metal Pot

In English plumbing circles not only a pot made of metal but also a melting pot for melting solder (see also METAL). In Scotland it is distinguished from the lead pot by 'solder pot'. Co Durham says 'metal pan'.

WC Pot

In Co Durham and the North Riding the term 'pot' means WC pan. There are the old-fashioned hopper pots, modern S-pots and P-pots ie pots with S-trap or P-trap (see TRAP). It is easy for an outsider to speak at cross purposes on the subject of pots and pans in that part of the country. (See also WATER CLOSET)

POT HEAD

In Somerset, a rainwater head presumably with a bow front or rounded front. It is rather coincidental that Somerset should call on a kitchen utensil on two occasions to provide names for rainwater heads (see also BUCKET HEAD). And it is perhaps coincidental that the name 'pot head' should also bear some meaning or relationship with the term 'soil-pipe head' (qv).

QUIRK

Heard in mid-Wales for a plugging chisel (qv). On architectural mouldings a narrow, sharp-edged groove in stonework is a quirk. A plugging chisel, or quirk, could be used to make such a groove.

RAGGLE

When one cuts out the cement joints preparatory to fitting step flashings, cuts a groove in brickwork or masonry to take the turn-in of raked flashings, or cuts a groove in a wall, the groove or raked out joint is called a raggle in Scotland. To raggle a chimney, say, is to cut a raggle, and one uses a raggling iron (qv).

In England and Wales one does not raggle, one chases, ie cuts a chase, or groove, in a wall, or one might rake out the joints in brickwork using a plugging chisel (qv).

Raglet

Buchan uses the term 'raglet', not raggle, and that could be a variation of 'reglet' which, in architecture, is not a groove but a narrow flat band used to separate mouldings or other parts from one another.

RAGGLING IRON

A Scottish term for a specially shaped chisel for raggling (see RAGGLE). It is best described as a cross-cut chisel, the blade being at right angles to the thinned part of the shank so that while cutting into a joint in brickwork, the blade can cut free of any binding by the stem. As the thicknesses of mortar joints vary, it is advisable to have raggling irons of two or three sizes. In England and Wales the plugging chisel (qv) is used to rake out joints, and could well be called a raggling iron.

RAINWATER HEAD

A vertical rainwater pipe of nominal bore can cope with an

enormous quantity of rainwater from roofs, but some roof gutters have cross-sections greater by far than any rainwater pipe and there are occasions when more than one gutter has to discharge into one rainwater pipe. A connecting piece with large open top to take gutter discharge, and a narrow bottom outlet to fit into a nominal sized rainwater pipe must be installed at the top of the rainwater stack. The connecting piece is known as the rainwater head, meaning, of course, a rainwater-pipe head.

Rainwater heads are not used in run of the mill Scottish plumbing; rainwater pipes are usually taken into branch

Fig 29   Hopper-type
          rainwater head

fittings and offsets for connecting to rones (qv), but throughout England and Wales they are used freely as heads, and as waste-heads (qv).

Shapes, sizes, materials used in making, and local terms, vary very much according to age and region. In some areas there are two or three terms used loosely to indicate *any* shape of head, and in other areas one local term, which might suggest a particular shape, is applied to all shapes. For example, hopper head (qv).

The intricate work of the old craftsmen is still to be found in rainwater heads which have been in use since medieval times. Such work is a reminder that the real plumber—the lead worker, not the pathetic gomerel presented to the public through the modern media—was indeed an artist, though he would be the first to ridicule the idea.

The rainwater heads once made of lead were of sheet lead bossed to the required shape, often highly ornamental. Some heads were a mixture of bossing and of lead casting, but so skilful was the bossing that it is difficult to determine without close inspection whether the lead is moulded by casting or by bossing from the sheet.

Not so common is the rainwater head of sheet copper, but when found it is recognisable by the lovely green which copper takes on with age and atmosphere.

There was great consternation in plumbing circles when cast-iron heads made their appearance near the end of the last century. It was felt that this was the beginning of the end of craftsmanship (which it was); that the craft could not bear this hard, unyielding new material. The dismay must have been akin to that of many plumbers who, in recent years, have watched the wave of plastics overwhelm what was left of the craft.

RAINWATER PIPE

A pipe made for the purpose of conveying rainwater from a roof. Cast-iron rainwater pipes have always come in lengths of up to 6ft; those of zinc in lengths of up to 8ft; PVC pipes now are made in lengths of up to 18ft. One length of pipe is a rainwater pipe, but when lengths are fixed, one on top of the other, to the face of a building, the result is still called 'a rainwater pipe', the word 'stack' being understood (see also STACK). The rainwater pipe need not always be vertical; sometimes a stretch of pipes is taken with only a slight fall to a point where it can branch into a vertical pipe; as the nearly horizontal pipe is not a stack, it is sometimes referred to as a 'run'.

Although rainwater pipe is made for the purpose of conveying rainwater—the joints being loose-fitting with no jointing material—it must be stated, with discredit to British Public Health authorities, that throughout the greater part of Britain the use of this pipe as waste pipe from sinks, baths and basins is permissible. A waste head (qv) may be inserted into a rainwater pipe stack and a sink waste, or bath waste etc taken into the waste head. The pipe above the waste head is, of course, a rainwater pipe, but the rainwater pipe below the waste head becomes, strictly speaking, a waste/rain pipe or rain/waste pipe—the combination of terms is irrelevant. The principle is bad.

RECEIVER

A straightforward word for a rainwater head in Aberystwyth; it receives rainwater or waste water.

REFLUX VALVE

A valve designed to prevent a reversal of flow. The action of the valve is automatic, being opened by the flow and closed by gravity when the flow stops or reverses. BS 4118 gives this as a preferred term. Flux, an early name for dysentery, means flow. Reflux means to flow back or return, so the term, reflux valve, would appear to mean a 'return valve'.

Non-return Valve

Although this term may suggest the opposite of 'reflux valve', it is applied to the same valve; indeed non-return valve, now a 'non-preferred' term, has long been used in the trade throughout Britain, and is better known than 'reflux valve'.

Back pressure Valve

The reflux valve comes into action by closing, when there is back pressure which stops the flow.

Check Valve

Another widely used term for a non-return valve and now, according to BS 4118, another non-preferred term for a reflux valve. (See also CLACK VALVE)

RESEALING TRAP

A trap so designed that after siphonic action or waving of water, caused by excessive pressure fluctuations at inlet or outlet, has taken place, a certain amount of water falls back into the trap to give sufficient seal. The well-known

Fig 30 McAlpine
resealing trap

McAlpine resealing trap is the simplest action to follow; when the seal has been broken the water held in the drum-shaped part falls back into the trap to give a seal.

Anti-siphon Trap

This term, which is frequently used, is usually a misnomer for 'resealing trap': McAlpine's resealing trap is often given this name. There has been much debating amongst plumbers as to whether 'anti-siphon' or 'resealing' is the correct description: the argument depends on what is meant by

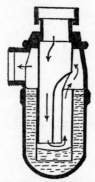

Fig 31 Grevak monitor

'anti'. If it means that the anti-siphon trap eliminates any tendency to siphon, then there are very few such traps used in Britain. The McAlpine Silentrap of recent years is probably an anti-siphon trap, as any negative pressure in the outlet is immediately recompensed by means of an air inlet valve on the crown of the trap. Greenwood & Hughes Ltd produce a very neat bottle trap, the 'Grevak Monitor', which they describe as an anti-siphon trap. The illustration shows the Grevak Monitor when subjected to severe siphonage. The air is drawn through the central by-pass tube leaving the water surrounding this tube to fall back and reseal the trap; the central tube breaks the siphon before an excess of water is lost and in that it might be described as anti-siphon, but still a certain amount of water must fall back to reseal it. Is it a resealing trap or an anti-siphon trap?

REVOLVERS

Used frequently in Scotland as a collective term for the primary flow and return, ie between boiler and hot water cylinder or tank, in a domestic hot water system: as the water completes a circuit, down return pipe and up flow pipe, it is said to revolve. The flow pipe is often referred to as the top revolve or top revolver and the return the bottom revolve or revolver.

RISER

Heard in Scotland meaning the flow pipe in a gravity hot water circulation (see also REVOLVERS). The water rises *from* the source heat *to* the point where it returns, in the return pipe, to the source of heat.

RISING MAIN

BS 4118 gives this term as *'deprecated'*, 'rising pipe' being the preferred term for a pipe in a building through which water rises from the ground level to a storage cistern. Deprecated or not, rising main, as well as riser, is used in many parts of Britain.

RONE

A half-round eaves gutter, but if the phrase 'half-round gutter' was used in the presence of a Scottish plumber, he would probably say: 'You must be joking'; to him it is a rone or rhone.

There has always been some controversy as to the spelling, rone or rhone. Perhaps the 'rhone' school of thought, with the Rhone Valley in mind, has an exaggerated conception of the French influence on Scottish language. In Scotland, as in other parts of Britain, a run is a stretch of running water, a small channel. Variations go from run to rin, ron, rone, and Chambers gives 'rone' and 'rhone', a small channel.

RONE BOLT

Rone bolts and nuts are used in Scotland in the joining of rones (qv) by means of spigot and faucet joints (qv). The bolts are usually $1\frac{1}{4}$ x $\frac{1}{4}$in with countersunk heads.

Gutter Bolt

In England, where the rone is known as a half-round eaves gutter, the Scottish rone bolt is called a gutter bolt. But the Scots, too, have gutter bolts; they are usually $\frac{5}{16}$in in diameter and are used in the joining of those ornamental gutters (qv) known in England as 'moulded gutters' (qv).

RONE PIPE

A rainwater pipe (qv). While the Scottish plumber calls rainwater pipes 'conductors', the Scottish layman speaks of 'rone pipes', ie the pipe coming down from the rone.

ROPER

In Scotland, where gaskin is commonly called rope or soft rope, the tool used for packing the rope into spigot and faucet joints is a roper or roping iron.

ROUND DRESSER

A dresser (qv) with a round cross-section, used in conjunction with other dressers or mallet in the bossing of sheet lead. Whereas the spoon dresser (qv) is for spooning thickness of lead from one part to another the round dresser is more for drawing lead up with solid strokes when forming external corners like those of a lead box. Many plumbers save themselves the expense of buying a round dresser by simply reversing the flat dresser and using the round handle as bossing requires.

The term is used throughout Scotland and the north of England down to the Midlands, where it is also a bossing dresser (see also BOSSING STICK).

RUNNER

A longscrew and, like longscrew, frequently used as a synonym of connector. A runner, as often as not called a running thread, is so called because it is made to allow a backnut and/or a socket to be turned freely, or to *run* freely, for the full length of the backnut and/or socket.

RUSSIAN TALLOW

Although Davies and Hellyer both mention the use of rushes with blowpipes, neither gives any detail about the rushes, but Davies mentions the grease from the rushes. It is reasonable to suppose that the rushes were, in fact, rush candles or tapers. Rush candles and tapers were made by dipping rushes into grease or tallow. (Thus, one 'dips one's wick' when performing the sex act.) Considering how the spoken word can be distorted when passed from generation to generation, from journeyman to apprentice, one is tempted to follow on from rush-and-tallow, to rush-an'-tallow—thus, 'Russian Tallow'.

However, tallow really was produced and exported from Russia on a large scale in the nineteenth century, and Russian tallow, a phrase still used in many parts of England and Wales, may well have reached plumbers' workshops. Many old plumbers swear that the *real* Russian tallow was the best.

SARKING

In Scottish dialect a sark is a shirt, a chemise, or a night-dress. (Cutty sark is part of a woman's underclothes cut short —a worthless woman). Sarking is a coarse linen shirting; the sawn boarding on rafters.

Many English plumbers know the word 'sarking' as being associated with special felt for lining roofs under slates and tiles, but have been unable to say why it should be called sarking; from the Scottish definition it is apparently shirting or covering.

The Scottish 'sarking', in the building trade, is sawn board-ing about five-eighths of an inch thick for covering the rafters before slating or tiling. In England and Wales the rafters are battened. The Scottish method of boarding the roof (roofing felt is laid between slates and sarking) is the reason for the great difference between the Scottish plumber's roofing methods and those of the south. That is good enough reason to mention 'sarking' in this work for plumbers.

The carpenter fixes his rafters, covers them with sarking and fixes the wooden ridge rolls on main ridges and piend ridges (hips). The roof is ready for the plumber. The plumber does his flashing of chimneys etc, fits expansion pipes (see SWAN NECK) and hangs rones or OGs. The roof is ready for the slater. When the slater has finished and before he takes down his scaffold boards, the plumber fixes the ridging —usually of zinc. There is no reason for one trade to be working on top of the other.

In England and Wales, where battened roofs are the fashion, the plumber does a certain amount of flashing, the slater does his part, then the plumber has to clamber over slates or tiles to carry on with the main part of the flashings.

SAUCER JOINT

As the making of a taft joint (qv) necessitates the flaring out or saucering, of one end of lead pipe, someone somewhere simply had to call it a saucer joint; that place was in the Llandello district of Wales.

SCREW

The plumber, like anyone else, uses this single word to mean wood screw, the 'wood' being understood and undeclared unless to differentiate between a wood screw and a self-tapping screw for metal. An innocent victim of metrication is the old rule passed from generation to generation for finding the gauge of a screw by measurement : measure the head in sixteenths, take one sixteenth away and double the remainder. A screw-head measuring $\frac{5}{16}$in would be a no 8 screw.

According to information received from GKN Screws & Fasteners Ltd at the end of 1970, the metrication of wood screws was such a complicated affair that the 'final dimensional form' had not been resolved and it was expected that metric wood screws would not be available in quantity for another four or five years.

A hint for making one-way screws for hinges and hasps of tool-boxes : file away diagonally opposite quarters of the screw head so that there is no gripping edge on the slot for turning it anti-clockwise.

Pipe Thread

When a plumber cuts a male thread on a pipe he threads it or screws it—pipe screwing is more common than pipe threading. But when the thread is female it is not screwed; it is tapped (see TAP).

In general terms, 'screw' has many meanings, conventional and otherwise; it is used in one unconventional sense in the rhyme which tells of The Day the Plumber Came :

He said he wouldn't be a squeak,
But he was here for half the week,
      The Plumber.
A drop of grease fell on the floor,
and when he burned his finger, swore,
and *screwed* the maid behind the door,
      The Plumber.

SEAL

The water in a trap (qv) which acts as a barrier to the passage of foul air through the trap, also known as 'water-seal' or 'trap-seal'. Buchan uses the phrase '. . . the drown or dip of the tongue' which might have been Scottish terms in Buchan's time—ie during the latter half of the nineteenth century.

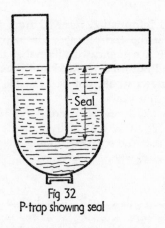

Fig 32
P-trap showing seal

SETTING-IN DRESSER

A dresser (qv) for driving, or setting, sheet lead into the angle of a gutter, say, at the side or back of a chimney. The cross-section is wedge-shaped or pear-shaped, the sharp edge being the setting-in part.

Setting-in-stick

From the Midlands to the south of England, this dresser becomes a stick—a setting-in stick.

Side Dresser

Describing a setting-in dresser, a plumber in Cornwall called it a side dresser.

SHANGIE

Grommet (qv). 'Shangie' comes to the Scottish plumber as easily as grommet or grummet. Chambers gives: 'Shangie, shanjie, shangy, . . . n . . . an ornament for a horse's tail; a shackle running on a stake to which a cow is bound in the byre; . . . a loop of gut or hide round the mast of a boat into which the lower end of the sprit is slipped . . .' (See also CRAMP-AND-RATCHET)

SHUTE HEAD

In North Devon, where eaves gutter (qv) is shuting, the rainwater head is a shute head.

SHUTING

Eaves gutters (qv). In Wiltshire, Devon and South Wales from Cardiff to St David's Head. Second-hand information gives the word as implying, in Somerset, half-round gutters only.

A middle-aged plumber in South Wales said: 'They keep on about those eaves gutters but I just write shuting in my time-sheet.'

Shutes, shuting

On ancient buildings, like cathedrals, these words are used for the lead channels, supported by boards, which nose out from the upper reaches of the building and shute the rainwater clear of the vertical stone faces.

I

SINK

In 1440, a sink was a pool or pit formed in the ground for the reception of waste water, sewage, etc. By the end of that century it was a conduit, drain or pipe for carrying away dirty water. 1566 saw it as a basin or receptacle made of stone, metal or wood, and having a pipe attached for carrying waste water to a drain. The original sink became known as a cesspool in 1671. The early stone kitchen sink was called a splash stone in some regions and slop stone in others (see also JAW-BOX). In Co Durham, a yard gully was referred to by elderly plumbers as a sink—the kitchen sink was also a sink.

Modern sinks made of perspex or any plastic material are poor substitutes for fireclay. The stainless steel sink is the fireclay sink's only true competitor, but stainless steel sinks too often are made without an overflow, and that can be a health hazard unless extreme care is taken in the fitting of waste pipes from them. Also, fireclay sinks, having flat bottoms, are slow emptiers; whereas many stainless steel sinks made with fast-emptying dished bottoms aggravate any tendency which the sink trap may have to siphon. The layman may have some sense of satisfaction when the stainless steel sink, particularly one without an overflow, empties quickly with a suck and gurgle, and he does not recognise those symptoms as the danger that they are.

Sinks, of course, are not made for the kitchen alone. BS 4118 gives a fine list of sinks for various purposes : bedpan sink or bedpan sluice, Belfast sink, bucket sink, cleaner's sink, combination sink, combination slop and wash-up sink, crockery sink, drip sink, Edinburgh sink, hospital sink, laboratory sink, London sink, mackintosh sink, plaster sink, pot sink or pot wash sink, pot washing sink or utensil sink.

SKELETON FLASHING

A term frequently heard with reference to the herringbone flashing, even as far north as Berwick-upon-Tweed where single steps are used.

SKELETON WASTE

A type of sink waste fitting used in conjunction with lead traps. The brass flange being soldered to the trap inlet, the strainer flange is bedded with paint and putty into the sink outlet, the long brass bolt passed down through it and screwed into the thread in the centre piece of the bottom flange which is also painted and puttied. The bottom flange is pulled up to the bottom of the sink to make a watertight joint. Sometimes the bottom flange, if the sink outlet is large, has insufficient grip on the sink and the best way to treat any such flange is to solder on a lead ring to enlarge the flange. The lead should be scored to form a key for paint and putty. Until recent years it was the fashion to use a lead ring on every bottom flange, but the practice has ceased, the result being many complaints of leaking sinks.

The skeleton waste is probably so called because, compared with the straight waste fitting with long thread, it is a mere skeleton. The old Scottish method, on the same principle, was even more of a skeleton (see BRIG-AND-NAIL)

In England and Wales the skeleton waste is known as a beacon waste or a Belfast waste (see also TABLE WASTE).

SKEW

The OED gives this as from the Old French *escu* which is a phonetic descendant of the Latin *scutum*, a shield : a stone specially intended or adapted for placing with other similar ones to form the sloping head or coping of a gable, rising slightly above the level of the roof—1533. The line of coping on a gable : chiefly Scottish.

To many Scots a skew is nothing more than that, whilst to many others it is a gutter between slates and the sloping head of a gable. Over the last hundred years the meaning has been extended, in many parts of Scotland, to the side gutter of a chimney, dormer, and the gutter necessary where a roof comes against the gable of a higher building. As well as the gutter, there are of course skew flashings, eg step flashings.

Secret Gutter

Misinformed English plumbers are sometimes under the impression that the Scottish method of guttering the side of a chimney, etc, is in fact secret guttering, but that is not so. The skew gutter has, open to view, sufficient water way to ensure that there is no blocking with dust or leaves. It could easily be made a secret gutter by extending the edge of the slates further over the gutter almost to touch the vertical side of the gutter.

English and Welsh plumbers may think it somewhat peculiar to speak in terms of skews and secret gutters at chimneys, but roofwork for the Scottish plumber is very different from the method of the south.

SLOP STONE

A name for the early stone sink (see SINK) which was carried on in some regions to describe the fireclay sink. In Scotland the term is still used in connection with dished tops for gulleys: dished gulley tops were once made of stone. With the advent of fireclay they were, and still are, made of that material. This slop stone has been referred to as a dish-stone in Scotland and Northumberland. BS 4118 gives 'dished top' and 'dish brick' as names for 'a shallow glazed stoneware receiver for setting on top of a gulley'.

SMUDGE

A mixture consisting chiefly of lampblack and glue size, which comes to the plumber nowadays as powder in packet, or cake in tin. When moistened with water the smudge takes the form of a thick water paint. Brushed on to the surface of metal around a point to be soldered, it tarnishes that surface and prevents the solder from tinning the metal out with the desired limits of the joint. Smudge can be kept moist or remoistened by working in with a smudge brush a few drops of water or, as is more usual, a spittle. Many old plumbers tell of the days when they made smudge with lampblack and stale beer, but Davies gives the most detailed instructions as to contents and as to the making:

Take a large packet of lampblack, put it into the metal pot and make it red hot, then let it cool. Take a lump of chalk about half the size of a large hen's egg, pound or break it up very fine or rub it on the rasp; then mix the chalk and the black together in water or beer, and well grind the mixture up, either with a muler and paint stone or with a trowel. Whichever it may be, the black must be well ground, *so that no grit can be found*. Make it as stiff as good mortar. Next, have some melted glue (here the glue-pot will be handy), put a good-sized table-spoonful with the black, warm up the lot over a moderate fire, keeping it well stirred up from the bottom. Then with an old worn sash-tool (paint-brush) free from paint or grease (which may be had of any painter, washed out) and well worked into the 'soil', paint or 'soil' (as it is termed) a piece of lead which is *perfectly* free from grease, and hold it to the fire to dry—not too close, so as to burn the soil. Then with the fingers rub it until it shines. If it rubs off, it wants more glue . . .

Along with the terms smudge and soil, others may be heard as alternatives throughout England and Wales like— plumber's black, black, lampblack, Burnham black, and tarnish. In Scotland, smudge is generally the one used.

Smudge, as a verb, means simply to apply smudge. (See also INITIATION)

SNAKE'S MOUTH

Given as an alternative to bird mouth (qv) in many parts of England, but in Wiltshire the ornamental end of a puff pipe is a snake's mouth. A puff pipe is a short length of pipe taken from the crown of a trap to ventilate it, to the exterior wall, ending at the face of the wall with an ornamentation supposed to represent a snake's mouth : a bead is turned on the end of the pipe, crossed wires soldered across the mouth to form a grid, and a short piece of copper wire sometimes soldered on as the snake's tongue.

SNOW BOX

A rainwater head is a snow box in parts of Somerset and Wales, although the meaning does not seem to relate to such a head. There seems to be some confusion of terms—a drip box (see CESSPOOL) has also been called a snow box.

SOAKER

BS 2717 : Flexible members, usually of metal lapped with slates, shingles or tiles and bent to form a watertight joint.

As far as plumber's flashings are concerned, the bent up part of the soaker is covered with the step or raked flashings. Although the soaker is the acknowledged best means of weathering in England and Wales, it is rarely used in Scotland (see SKEW). The soaker is bent ready for fitting; the upstand or vertical part being placed against the brickwork and covered by the flashing, while the other part is lapped in with the courses of slates or tiles.

A common method of weathering a chimney or the like in England and Wales is to fill in the angle between brickwork and slates with a cement mortar fillet which covers the soakers, and that can be quite effective when properly executed. However, such a method is open to abuse by cheap-jack builders and handymen who form a neat fillet of cement mortar which is pleasing to the clients' eyes, but omit the soakers.

On new work, the roofer fits the soakers which the plumber provides cut and bent to the roofer's requirement.

Plumbers in Hawick referred to step flashings as soakers, but the misnomer is unusual.

SOCKET

The English and Welsh term for the Scottish faucet (qv). But the word has two meanings in plumbing; it is also the BS preferred term for a coupler (qv).

SOIL PIPE HEAD

Less than a hundred years ago the rainwater head (qv) was also used as a soil pipe head, just as today it is used as a waste pipe head (see WASTE HEAD). The day may come when the waste head is considered with much the same revulsion as we now feel towards the soil pipe head of yore. This soil pipe system was known as Norman Shaw's system;

it comprised a trapless water closet discharging into an outside head on a soil pipe which in turn discharged into a primitive trap at the bottom of the stack.

Richard Norman Shaw (1831–1912) was an architect of distinction who, in 1854, won the Academy Gold Medal in architecture—not in plumbing! But it is easy to be scornful, and it is a fact that the problem of combining good plumbing with aesthetic architecture is very, very often a problem which is incapable of solution. Either the demands of good plumbing destroy the effect the architecture was meant to have, or architecture wins at the expense of the plumbing. Norman Shaw was a great architect and a mighty bad plumber.

Because of the old association of buckets and pots with closets, dry and otherwise, the terms bucket head (qv) and pot head (qv) could well have originated in the Norman

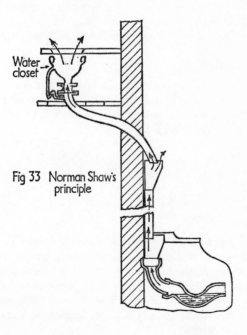

Fig 33 Norman Shaw's principle

Shaw soil system. The maintenance plumber of today must shudder at the nightmarish thought of clearing a blocked soil pipe head in the old days. Hellyer indicates with arrows (in a sketch) the stream of foul air from trap to head and then through the trapless closet.

SOLDER

BS 4118 defines the term with an admirable economy of words: 'An alloy used in a molten state for joining certain metals together.'

The solder used by the traditional plumber is a soft solder, ie an alloy of lead and tin; hard solder consists of copper and small amounts of other metals and is used in brazing and bronze welding.

As a verb, 'solder' means to join metals together with solder. But verb or noun, the term becomes souder or sowder, sooder and even sodder, in the dialects of Britain, particularly in northern parts of England and in Scotland.

Coarse Solder

The name given by Scottish plumbers to solder used in making wiped joints and wiped seams, as opposed to a fine solder. Such solder consists of two parts of lead to one part of tin (see also METAL)

Fine Solder

Consists of roughly equal parts of lead and tin. It is used in conjunction with the copper bolt or with the blowpipe lamp.

Bar, Stick, Strap

Whereas the English or Welsh plumber speaks of a stick of metal (qv), the Scottish plumber would say: 'a bar of solder'; to the Scottish plumber the bar is coarse or jointing solder, and the stick fine solder. Buchan refers to fine solder as strap solder. The English and Welsh plumbers when requiring fine solder would ask for 'a stick of tinman's' or '. . . tinman's solder'.

SPICKET

As well as meaning the spigot (qv) of a pipe, the Scottish spigot or spicket (mainly in the vernacular) is a tap. Once used with reference to the wooden tap for a cask, it has been extended to take in the tap over the kitchen sink or any bib tap.

The OED says of spigot that it was late ME (probably from Provencal *espigot* formed on *espiga*, spike). It was a small wooden peg or pin, once called a spile peg, for stopping the vent-hole of a barrel, or a similar peg inserted into the opening or tube of a faucet for regulating the flow of liquid, then about the beginning of the eighteenth century it was synonymous with faucet which at that time was a wooden tap. It is interesting that when 'spigot' and 'faucet' were dropped from the English language, with the meaning 'tap', the Americans adopted 'faucet' and the Scots 'spigot', with their 'spicket'. The original spigot was, of course, a cock (qv) in the true sense of the word and not a screw-down type like those of today.

SPIGOT

The plain end of a pipe or gutter which is inserted into the faucet (qv) of the next section for the formation of a spigot and faucet joint (qv). With certain cast iron pipes this plain end is made with a thickening which helps to centre the spigot in the faucet of the next length.

In Scottish dialect spigot becomes spicket (qv).

SPIGOT AND FAUCET JOINT

Or spigot and socket (see SPIGOT and also FAUCET). A joint in which the spigot of one section of pipe or gutter fits into the faucet or socket of the next section. The annular space between spigot and faucet may be filled with jointing material such as lead, cement, putty or some patent material. Spigot and faucet joints on rainwater pipes should not be sealed, so that should there be a blockage in the pipe the water will escape out of the first joint above the blockage; if the joints are sealed, the water held back by the blockage rises to the top of the stack and spoils the face of the building when the pipehead subsequently overflows.

Sir Thomas Simpson is said to have invented the spigot and faucet joint in cast-iron pipes while he was employed as an engineer with the Chelsea Water Company in London in 1885. It is interesting that the earliest date given by OED for the term is 1897, just a few years after its invention. But Simpson's invention was not what one might call an original idea; before that date such a joint used for wooden pipes was called a spigot joint. In Hawick Museum, Scotland, a wooden pipe, taken out of an old well, can be seen with one end of the pipe tapered and the other end scooped out to receive the spigot of the next length.

Cock-and-pail

This is given by Chambers as a Scottish phrase for spigot and faucet, but this Scottish writer has never heard it. Obviously the cock is the spigot or male end and the pail, being the faucet or female end, receives the cock.

SPOON DRESSER

To spoon is to lift something with a spoon, or with a spooning motion. Sheet lead can be dressed in such a way that the thickness can be taken from one part to another, by spooning, as required in bossing sheet lead or in pipe bending. In pipe bending, for example, the lead tends to thicken at the throat and sides of the intended bend and thin at the heel, so as bending proceeds, the spoon dresser is constantly used to dress the thickness from the throat and sides to the heel; therein lies the art of pipe bending.

'Oval dresser' or 'spoon dresser' are terms used in Scotland and the northern half of England. Travel south from the Midlands and the name changes to bending dresser, bending spoon, spoon stick, and then, in London and the south east, and from Kent to Cornwall, one hears 'bending stick'. The 'bending' part of those terms refers to the use of the spoon when bending lead pipes, but the spoon dresser is used on innumerable occasions where thickness of lead has to be moved from one part of the sheet to the other.

SPOUT HEAD

In areas where eaves gutters are spouts or spouting, and where rainwater pipes are down spouts, one may expect to hear 'spout head' for rainwater head.

SPOUTING

Eaves gutters (qv) in various areas in Yorkshire, from West Riding to North Riding.

Spouts

'Spouting' becomes 'spouts' in Northumberland, Westmorland and Cumberland.

Spoots, Spootin'

Just north of the Anglo-Scottish border, in Langham, eaves gutters as well as being rones (qv) are spoots or spootin', dialect variations of spouting. However, in some parts of the above areas the spout or spouting is the nozzle outlet of a rone.

STACK

A term which requires a qualifying word unless the meaning is understood in context of conversation. In the flashing context a plumber flashes the chimney or he flashes the stack. Also, there are rainwater stacks, waste stacks, soil stacks, and vent stacks. A vent stack is a pipe stack which serves only as a vent for the drain, and should not be confused with a stack vent which is a vent for a stack—ie for a soil stack or a waste stack.

STACK HEAD

A rainwater head. Used in conversation with plumbers in

random parts of southern England. It is not necessarily a local term as men who have used it are men of parts, so they could have picked it up anywhere along the way.

STEMMER

In mining, stem means to plug or tamp a hole for blasting and a stemmer is a metal bar used for stemming.

The word has been heard in Co Durham from a retired plumber, with reference to the tool for packing yarn (see GASKIN) into a spigot and socket joint.

STEP FLASHING

BS 2717 : 'A flashing used to cover an inclined intersection, its upper edge being shaped to step up from course to course of brickwork or masonry and secured into the horizontal joints.'

Flashings such as these, usually called step flashings, cannot be defined in a matter of thirty words; one can have nothing but sympathy for a compiler who may have been forced to do so within those limits. There are two distinct shapes or steps and at least five different ways in which they may be used. Some methods do indeed cover inclined intersections, but others merely cover the upstands of gutters or of soakers which cover the intersection. In the case of masonry, due to the irregularity in sizes of stones, one separate step could vary so much in size from another that step flashings which covered the intersection would be uneconomical and difficult to contrive. Methods of flashing with steps are known as : (1) the step and cover flashing; (2) the combined step and gutter; (3) the hanging step flashing with soakers; (4) the hanging separate step flashing and soakers; (5) the hanging separate step flashing and gutter.

The gutters may be secret or like the Scottish skew.

The first two methods would permit covering the intersection and shaping the steps out of one strip of lead. In the other three cases the step flashing is used in conjunction with other components which cover the inclined intersection. A textbook containing all technical details might be difficult to find in these days of standardisation; the methods of Scots, Welsh and English have seldom been detailed in one textbook.

Davies described the two different shapes as 'the original or real step flashing' and 'the real herringbone flashing'; he claimed that the latter was not a *stepped* flashing.

The Real Step

This is in the shape of a step—vertical riser and horizontal tread—and is used in such a way that, starting at the lowest point, each step laps over the next. The common terms are 'separate step flashing' or 'single step flashing'.

It has been said (in London) that the separate step method is practised in northern districts. 'Up North' can mean any place north of Watford when one is in London, but even at that the statement about where steps are used is inaccurate. Travel to the West Country—there are single step flashings from South Cornwall to North Devon. Again, travelling west from Herefordshire, as one penetrates Wales the herringbone flashing becomes less apparent and, once in Wales proper, the single step is general. On the other side of England, the herringbone is the practice throughout Kent, then north through East Anglia to Lincolnshire. In Lincolnshire the herringbone is used on houses of brick but those of stone usually have single steps. In the East Riding of Yorkshire, houses built in recent years have herringbone, while others have single steps. As one progresses north of a line from Lincoln to Chester, the herringbone becomes less frequent, and in Northumberland, Cumberland, places between, and throughout Scotland, single step flashing is the fashion.

While Davies was referring to this flashing as 'the original step flashing', Buchan was calling it 'a stepped apron'. (See APRON)

Herring Bone

Usually referred to as a 'step flashing'. Lengths of this flashing can be cut from one strip of lead, zinc, etc and, as

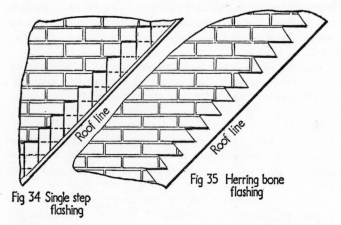

Fig 34 Single step
flashing

Fig 35 Herring bone
flashing

there is not any lapping as with single step flashing, it is
said to be more economical. There is little doubt that it will
become the standard flashing in time, and every house in
Britain will conform to the standard. The main cause of this
evolution can be put down to modern plumbing practice.
Not many years ago it was the custom for plumbing firms
to buy sheet lead by the sheet—ie, in rolls which, when rolled
out, measured about thirty feet by eight feet, and plumbers
had to cut lead as required and in such a way as to waste as
little as possible. Any odd pieces or off-cuts could be used to
make single steps. Nowadays, makers supply sheet lead in
strip on reels and the plumber orders to the required width :
there is very little waste—not enough to make single steps,
so the modern method of supply is most suitable for herringbone flashing. Zinc sheet is the favourite for herringbone as
it can be cut to shape in long lengths and retain some
rigidity and the turn-in is easily made with a step turner—
a special tool well known 'in the South' but practically un-
known 'in the North'.

STOPCOCK

BS 4118 says of a stopcock : 'A cock fitted in a pipeline
for regulating the flow of water' (see also STOPVALVE).

Stopcocks, ie the cock type of stopvalve, are used through-out Scotland, in Northumberland, the North Riding, and North Devon, and probably many other places, as under-ground stopcocks. In recent years the water authority in South Cornwall has removed the old stopcocks and replaced them with screwdown stopvalves, while in North Devon the screwdowns have been removed and replaced with new cock-type. But in Somerset, when asked about main stopcocks, an unblushing representative of the water authority cried: 'Cocks are out!' He meant, of course, that they use screw-downs and not cocks, the reason being that, owing to the extreme hardness of the water, cocks would lime in a short time and the metal plug in the cock become immovable.

STOPVALVE

A valve (qv) fitted in a pipeline for controlling the flow of water, the type mostly used by plumbers (not heating engineers) being the screwdown, which has a disc, usually with a washer, component adjusted by a worm or screw; the screwdown stopvalve was invented by Edward Chrimes of Rotherham in 1845, thus the name 'Rotherham valve' much used in the latter part of the nineteenth century. BS 4118 gives 'stop tap' as a non-preferred term and rightly so (see TAP).

Elderly plumbers sometimes use terms like 'under and over' and 'up and over' to distinguish the screwdown valve from the plug type of stopvalve—ie stopcock; when the disc is raised the water comes up from *under* the disc and passes *over* into the continuing pipe-line. In other words, the water passes *up* through the annular seating and *over* into the con-tinuing pipe-line.

Stopcock (see also STOPCOCK)

Used very loosely in most parts of Britain, whether the valve is indeed a cock or not. The term screwdown stop-cock is often meant to indicate that the valve is of the Rotherham type and not really a cock.

UG stopcock

Often referred to as 'the UG' in Scotland, meaning underground stopcock. It is the stopvalve on the water service to properties. In Scotland it is more often than not under the highway. The name of this stopvalve is more abused than any other; no one seems to have decided that 'stopvalve' is the most suitable as it applies to valves whether they are plug-type or Rotherham. In England, most of those which plumbers call stopcocks or stop taps are not cocks. Some water authorities have 'STOP TAP' imprinted on their surface boxes but other authorities avoid the issue with the imprint 'W' for water. Newcastle & Gateshead Water Company called them stop taps in 1963 when the company changed hands, since which time the official term has been 'stopcock'.

STREET ELBOW

An elbow (qv) with male thread on one end and female on the other, also referred to in England and Wales as an M and F elbow. The term is not known in Scotland.

It refers to M and F beaded elbows only, that is, those with an integral strengthening bead. (Any unbeaded fitting is known as 'plain'—eg plain elbow, plain tee, etc.) A National Price List refers to '. . . Street M x F' as if it could be a street elbow if it was not M x F.

'Street elbow' was once very much used particularly in the south, but it is giving way to 'M and F elbow'.

As to the origin of the term 'street elbow', a representative of Crane Ltd has been the only person to offer an explanation. It probably originated in the United States of America, where it is still used. When making a connection to a street gas main or water main, by boring and tapping on the top side of the main, a male and female elbow is screwed into the new tapping, the male end into the main and the female end turned to receive the male end of the horizontal service pipe. As such connections were made in the street mains, the elbow became known as a street elbow. If that suggestion is correct, it is strange that Messrs Kay Ltd's catalogue of pipe fittings should illustrate a 'street bend' for copper tube; such a bend would not be used for a street connection. (See also US STREIT ELBOW)

STUD

A running nipple. A very short running nipple is occasionally referred to as a stub nipple, but 'stub' is probably a variation of 'stud'. In various parts of the British Isles studding or stud nipple is a term describing a running nipple of small bore used in conjunction with other components, like a backplate and pipe-ring in the making of pipe brackets or hangers.

SWAN-NECK

Used to describe bends which in many cases bear little resemblance to a swan's neck (see OFFSET).

The long, bent outlet of a sink mixing tap is often called a swan-neck; the old-fashioned tube-crane, once very common in Scotland, was called a swan-neck and it *was* in the shape of a swan's neck. Gas wall-brackets are often called swan-necks.

In Scotland, where expansion pipes from domestic hot water systems are taken through roofs by means of lead slates and do not discharge over storage cisterns as in England, the plumber bends the top part of the expansion pipe to the shape of a swan-neck; it is referred to as a swan-neck.

TABLE WASTE

A type of waste fitting used in Scotland around the early 1920s, quite similar to the skeleton waste (qv) common in England and Wales in recent years.

TAFT

Usually a verb signifying the opening out of the end of a
K

lead pipe and turning over the edge to form a lip at right angles to the pipe, but is sometimes used to mean tapping lightly to move a thickness of lead. The word is little known in England and Wales, having fallen out of use, but in a catalogue sent out by McAlpine Ltd of Glasgow, makers of lead traps, lead soil pieces, etc, in 1970, 'tafted over' is used in describing lead soil pieces.

Gutter Stop End

In Norfolk 'taft' is used as a noun for a stop end in cast-iron gutter. The name may be from earlier times when plumbers tafted sheet lead to fit the ends of iron gutters; that kind of tafting would be more like bossing lightly with a mallet.

TAFT JOINT

A solder joint on lead pipe involving the tafting out of one end of the pipe to be soldered. Nearly every plumber in the land is familiar with the term but those who do not know taft, do not know why the joint should have this name; an elderly plumber in Wales thought 'taffy-joint' was the right name because, he claimed, that method of jointing originated in Wales. (See also TAG JOINT)

TAG JOINT

A taft joint (qv) requires little skill in the making and is often used for adding, or tagging on, short lengthening pieces to existing pipes and a bastardised form of taft joint (see CUP JOINT) is often made by incompetent plumbers when tagging on brass couplings to lead pipes. The term 'tag joint' was heard in Cornwall, but it isn't clear if the tag is a proper taft joint, or the bastardised form, or both.

TAMPION OR TOMPION

A turnpin (qv). The OED gives 'tompion' as a variation of

'tampion' (from French *tampon,* nasalized variation of French *tapon,* derivative of *tape* meaning plug). In English, 'tampon' means a plug inserted tightly into a wound or orifice. During the last quarter of the nineteenth century, while Hellyer and Davies were writing of turnpins, their Scottish contemporary, Buchan, wrote of a 'tompion or pipe widener'. As far back as 1625 a tompion, in general terms, was a plug of wood inserted into the muzzle of a gun to keep out rain, sea-water, etc. Since Buchan's time the word, in turnpin sense, has fallen out of use but its variations and its meaning remain.

Tampin

A variation of tampion heard in many parts of Britain, particularly from the not-so-young.

Tanpin

Had turnpins been made of oak, there might have been good reason to suppose that 'tanpin' is from Breton *tann,* meaning oak. But they are of box, so it may be assumed that tanpin is a variation, perhaps a corruption of tampin. It may be heard in many random parts of Britain and seen in a list of plumber's tools issued by the National Federation of Plumbers and Domestic Heating Engineers.

Tumkin, Tunkin or Tungkin

Scottish variations in the spoken word, any of which could be a compromise between turnpin, tampin and tanpin— or even tompion. A few Scots write their particular variation phonetically, but most, in the written word, would give turnpin, tanpin, or tampin in that order of preference.

The Scots readily convert their nouns to verb or adjective; the plumber 'tungkins' the end of a lead pipe and the resulting 'tunkin'd end' is an everyday term meaning a socket or bellmouth on a lead pipe. (See also DOOK)

TAP

Generally, used in many senses all with the original meaning of tapping a cask—tapping a source of supply or turning on a tap; borrowing. Partridge taps Grose as a source for '*tap a girl,* to deflower her'.

In plumbing and allied trades, one taps a female thread with a tool called a tap. The tap consists of a male thread on a tapered tool of hardened steel, grooved lengthwise to form cutting edges, and having a square head on to which may be fitted a ratchet for turning the tap. Tap, in this case, probably stems from the original meaning, a tapering cylindrical stick or peg, like a tap root.

A tapering peg, called a tap, was used for closing the vent hole in a cask (see also SPICKET); hence a hollow tubular plug through which liquid could be drawn, with a device for controlling the flow; a cock (qv) or faucet (qv). (See also VALVE)

TAPERING COCK

Heard in mid-Wales for a plug-cock. The descriptive word apparently refers to the shape of the seating.

THIMBLE

The traditional thimble, in England and Wales, is a brass pipe-fitting consisting of a spigot (qv) which is joined to a lead waste or a lead soil pipe by means of a solder joint, and a socket (qv) designed to receive the outlet of a WC pan or slop hopper, or the spigot of a cast-iron pipe. The thimble may be straight or bent. In various places it may be called a 'brass cup' or a 'brass socket'. Makers of copper pipe fittings now supply a similar cupped fitting with, in place of a spigot, an end designed for connection to copper pipe. Thimbles are seldom used in Scotland, where it is more usual in the case of WC outlets for the plumber to form a socket on the end of lead pipe by driving a bobbin into the end of the pipe with a dolly or by bossing out with a dummy.

THUMB BAT

A short wrought iron wedge (a flashing hook, qv) turned over at right angles near the wide end for driving into mortar joints in brickwork when fixing lead flashings. Just as 'bat' (qv) is more likely to be heard in Scotland and northern parts of England, so is 'thumb bat'. As bats are usually made out of odd scrap cuttings of lead, it would be difficult to say whether the bat or the thumb bat is the cheaper, but one thing is certain, a carelessly driven thumb bat can more easily cut into and perforate the vital part of a flashing with its gripping edge.

TILTING FILLET

BS 2717 has 'tilting piece'. A fillet of wood, its end section being a right-angled triangle, which comes in random lengths that can be cut as required. It is nailed to the roof under the outside edges of lead gutters, like valleys, side gutters and back gutters, and the lead is dressed over it. It is used to support the slates or tiles in the correct position relative to the roof surface. (See Figs 8 and 16)

TINKER'S FLEA

In South Cornwall, the sting of a spot of hot solder on the hand or face.

TINMAN'S SOLDER

Also known as tinsmith's solder, this term and its variant are more likely to be heard in England and Wales, but seldom in Scotland where it would be called fine solder (see SOLDER).

TOBY

In Scotland, where the main underground stopcock (qv) is usually in the road, what might be called a heavy duty stopcock box is used to withstand wear and tear by traffic. This

box is a toby, so called, perhaps, because of its close association with the highway or high-toby.

Partridge gives one meaning of 'toby' as a lady's collar,
about 1882–1918, from the frilly collar worn by 'Dog Toby'
of 'Punch and Judy' and suggests that there might be some
connection between that collar and a steel helmet called a
toby in the First World War. It could be that the Scottish
street toby with its bottom flange was from the hat that Toby
wore. However, whatever the source, one is reminded of the
thousands of Scots, including very many plumbers, all
familiar with the street toby, who were in the First World
War, and it is certain that Scottish wit would christen the
steel helmet 'toby'.

Top Hat

Water engineers in North Cornwall and North Devon, while
not familiar with 'toby', insinuate the same meaning when
they call their surface boxes 'top hats'.

Cock Case

In Northumberland 'toby' is used as in Scotland and until
1963 it was a standard term with Newcastle and Gateshead
Water Company. In 1963 the company changed hands and
the new body decreed that 'toby' would be dropped and from
that time onward 'cock case' was established as the official
name.

'Toby' is given by Partridge as slang for the buttocks or the
female pudendum, and one wonders if Newcastle and Gateshead Water Company, knowing the vulgar connotations of
the word toby, banned it for a more proper cock case.

TOBY KEY

A Scottish term meaning what is known throughout Britain
as either stopcock key, turn key, or water key. To reach the
stopcock (qv) the plumber lifts the toby lid and puts his toby
key into the toby (qv). In Northumberland and the south
of Scotland, where toby is a trade word, the term is not
known.

In Scotland, the term is occasionally used in workshops as a crude reference to the penis.

TOMMY

A podger (qv). Also known as a Tommy bar. Davies referred to the bent bolt (qv) as a Tommy, although that tool, in those days, was not so much the lever it has become in recent years.

TOMMY IRON

A plumber's iron (qv) in South Wales, as heard from an elderly gentleman in Pontypool. 'Tommy' is used in various trades as meaning an iron bar (see PODGER) and this is a likely term, as the plumber's iron was, in part, an iron bar.

TONGUE STICK

A spoon dresser (qv). When shown a spoon dresser, an elderly gentleman in South Wales called it a tongue stick or tongue dresser, and he held it upside down to prove that it looked like a large tongue. It did.

TOUCH

This old name for tallow is still used by many plumbers in England and Wales. 'Touch' is the flux for soldering lead with plumbers' metal and in the days of jack pumps a mixture of touch and hot resin was used to cement in the sucker box at the bottom of the lead barrel. Before the Second World War, in the 1930s, conceited apprentices would touch their hair with tallow to lend the coiffeur that desirable film star sleekness.

Davies says: 'Touch. Common tallow candle', which leaves one in doubt as to whether the tallow or the candle is touch (see also PLUMBER'S CANDLE). However, he also tells that a touch box is a tin box to hold tallow, suggesting that he really knows it is the tallow that is touch.

The OED says of the word 'tallow' that it is from late Middle English *talgh*, corresponding with Middle Low German *talg, talch*. . . . Chambers gives talgh, tauch, and taugh, so it could be that the English touch is from the late ME *talgh* or it could be a corruption of the Scottish tauch.

Grease

'Touch' is not used in Scottish plumbing jargon; 'tallow' is sometimes used but the everyday word is grease; pronounced in typical Scottish fashion with hard g, trilled r and s as z— greeze. Buchan writes of grease from tallow candle. Also, whereas the English plumber may have his touch box, the Scot has his grease box.

TRAP

BS 4118 : 'A fitting or part of an appliance or pipe arranged to retain water so as to prevent the passage of foul air.' (That is in sanitation : a fitting called a steam trap is used in steam heating systems.)

Nowadays, amongst plumbers in Britain, there is no alternative for 'trap', but the layman frequently refers to the 'trap under the kitchen sink' or under the wash basin as 'the S-bend' or 'the U-bend', and Davies in his day called it 'stink trap' and 'water lute'; many of the early traps did indeed stink, and thus the stink trap was well named, but Davies did not disclose why the trap should be called a water lute. One can only guess that the latter term came from the lute-like appearance of the old D-trap (see P-trap below). So many traps have been invented over the last two hundred years that this subject alone could fill a book and still leave a few traps unmentioned. After naming about thirty-six varieties Davies says, almost nonchalantly, 'The above are only a few quoted from memory . . .' BS 4118 gives eighteen types of traps but there are many varieties not mentioned of some of those types.

The multitude of traps invented over the years is due in

part to the fact that attempts have been made to shape them
for use in particular situations, but mainly because inventive
minds have been, and still are, bent on producing a trap
which is siphon-proof under every condition. Tube traps like
the P, the Q, and the S (see below) were called siphon-traps
during the last century and at the beginning of this one
because their very design induced siphonic action when water
was discharged from an appliance.

P-trap

Until the turn of the century this was called also '$\frac{1}{2}$S-trap'
(see also US P-TRAP). Davies insisted that $\frac{1}{2}$S was the
correct name for this tube trap and that the trap which most
plumbers called a D-trap was a P. The D-trap was obsolete
by the end of the nineteenth century but it is interesting that
it was used as far back as the seventeenth century; Davies

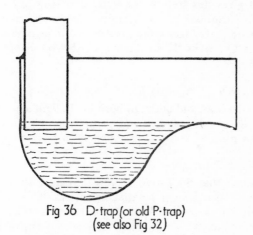

Fig 36   D-trap (or old P-trap)
(see also Fig 32)

gives an illustration of a D-trap or P-trap, call it what he
will, which he took out of Lothbury Old Church, near the
Bank of England, and this trap had been in use for over two
hundred years, bearing the initials and date 'T.L. 1678'.

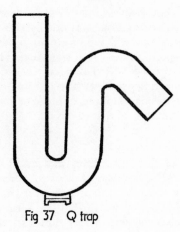

Fig 37   Q trap

Q-trap

BS 4118 gives this trap as obsolescent. It is as often as not called a ¾S-trap but in South Wales a few plumbers refer to it as a K-trap. It takes quite a stretch of imagination to see the letter 'Q', 'q' or 'K' in this shape; ¾S is probably nearer the mark.

Siphon trap

This has the inlet and outlet in horizontal alignment. Makers of drainage goods call it a siphon trap but plumbers refer to it as a running trap or, occasionally, a belly trap.

S-trap

Davies states that the S-trap is properly called a siphon and has always been known by that name, and to support his statement he says that Apollodorus called a tube of this shape a 'sipho'. This particular Apollodorus (there were a few others before him) was a Greek architect in the second century AD and, allowing for the fact that he was an architect who didn't know much about plumbing, it is very likely that he did indeed use the word 'sipho' as would any other Greek of the time: *sipho*, in Latin and in Greek, simply means a pipe or tube.

But it is true that the S-trap will induce siphonic action at the least provocation when it receives a discharge from an appliance. British public health authorities would do well to prohibit the use of the S-trap as does the USA National Plumbing Code.

The first S-trap to appear on patent specification was Cumming's water closet of 1775 and it is noticeable that he did not give it a name although he referred to 'the stink trap hitherto used for waterclosets . . .' as being a magazine of foetid matter; he probably meant the D-trap. Cumming's drawing (see WATER CLOSET) shows a perfectly good S-trap which he describes as a pipe 'recurved about twelve or eighteen inches below the pan or bason . . .' Although he dealt with this 'recurved pipe' as if it was his own idea and therefore completely new, it must be remembered that he was a watchmaker and could have picked up a few tips when looking over some plumber's shoulder.

TRAUING, TROWING

Variations of troughing in the dialects of the valleys of South Wales. When questioned about the spelling of their words, elderly plumbers gave t-r-o-u-g-h-i-n-g and added that 'troughing' was the correct word. Even the Welsh try 'proper English'.

TROUGHING

A self-explanatory word for eaves gutters (qv) which may be heard in many places in a line from Suffolk to Hereford and South Wales.

TRUMPET

A street elbow (qv). An elderly plumber in Cockermouth, Cumberland, identified the street elbow as a trumpet; another old gentleman in the same area held one to his mouth to demonstrate why it could be called a trumpet.

TUB AND SINK

A sink of nominal depth and a deeper washtub fitted side by side to make one unit; one waste-trap only is required. The washtub and sink may be made in one piece. The combination tub and sink, with central wringer board, has been fitted in most Scottish housing schemes for many years, the unit being of fireclay. (See also US TRAY)

TUMBLER-COCK

When an elderly gentleman in South Cornwall was asked about the type of stopcocks used there, he described them as high pressure screw-down cocks and added that as they were sent all the way from London, they must be good. On the question of plug cocks he said : 'Oh, you mean the old tumbler-cock'. The tumbler-cock or plug cock (qv) was once used in South Cornwall and has been replaced by Rotherham-type valves (see VALVE).

The tumbler-cock is so called because of the taper, like a tumbler, of the seating.

TURNCOCK

In the early years of the nineteenth century, when the inadequate water mains in the streets of London could supply water at low pressure only, it was the turncock, an employee of the water company, who had the business of opening at proper intervals the communicating cocks to small areas, each in turn. The frequency of this intermittent supply to an area varied under different circumstances, but the common practice in London was to supply water for from one to two hours daily, but not on Sundays; long enough for the storage cisterns to fill.

As water mains became established in other large towns and cities, turncocks were necessary to the system. The term 'turncock' is still used in parts of England, and it is still his job to be familiar with the water mains and valves in his district and to open and close valves as necessary.

TURNPIN

A cone-shaped, hardwood tool, usually of box, for belling out the end of a lead pipe to form a socket which receives the corresponding candle (qv) or male end preparatory to jointing. The turnpin should be wetted with water or spit to prevent sticking while being tapped into the end of the pipe with mallet or small hammer and it should be turned frequently during the operation. Possibly it is so called because of the turning motion while in use, but it is more likely that the name is due to the manufacturing process, which is mainly one of turning in a lathe.

The term is more common in England and Wales than in Scotland, but many plumbers, not allowing that a thing can have more than one correct name, debate whether the *right* one is turnpin, tampin, or tanpin (see TAMPION).

UNDERHAND JOINT

The soldered joint used to join two horizontal lead pipes; when the pipes are vertical an 'upright joint' is made. 'Underhand joint' may occasionally be heard in Scotland but it is mainly an English or Welsh term; plumbers in Scotland and parts of English counties bordering Scotland refer to it as a ball joint, a round joint or a straight joint. When a ball joint or round joint is made in an upright position it is usually called an upright joint as in England.

Plumbers in the south of England, particularly in London and the south east, tend to ridicule the Scottish ball joint which is much shorter than the English style of joint, and the traditional canniness or meanness of the Scots is often cited over the ball joint, which is said to require less solder than an English plumber would use in his style of joint. Regardless of common sense, the plumbers in the south, again particularly in the London area, follow a rigid rule regarding lengths of solder joints; textbooks have long stuck to a table of lengths : 3in length for a joint on a $\frac{1}{2}$in or $\frac{3}{4}$in water pipe, $3\frac{1}{4}$in length on $1\frac{1}{4}$in pipe, and so on.

A Scottish plumber might make his joints slightly longer

or shorter than another Scottish plumber; one joint may be 2½in and another 2¾ depending on the plumber. A plumber with a short handspan tends to make a shorter joint than a plumber with a wide handspan.

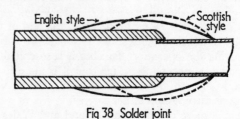

Fig 38 Solder joint

The lack of imagination in adhering to a fixed length of joint is best illustrated in the circumstance where a short brass coupling, say, has to be joined to a lead pipe. The English plumber sticks to his rule—the Scottish plumber makes his joint shorter than usual. The strength of the slim English joint is at the centre, where it is not wanted in this case. Countless such joints have been found pulled apart in London, especially on hot water pipes, owing to the extreme hot and cold such a pipe is subject to, and in times of frost this type of joint, being the weakest part of the pipe, breaks apart when frozen.

VALVE

BS 4118 cleverly defines a valve : 'A device for controlling the flow of a fluid, having an aperture which can be wholly or partially closed by the movement relative to a seating of a component in the form of a plate or disk, a door or gate, a piston, a plug or a ball.'

That definition covers just about every means of controlling the flow of fluid.

Cock

As defined by the same BS, a cock is apparently a valve in which the component is a plug : 'A device for controlling the

flow of water, comprising a body having a parallel or taper
seating into which is fitted a rotatable plug with a waterway which can be displaced relative to the waterway through
the body.'

But a valve is not always a cock, although plumbers
throughout Britain use the word 'cock' (eg see STOPCOCK
and BALLCOCK) to describe a valve of the type which has
a disc or a piston as the component (see also TAP)

Tap

A definition in the above BS gives 'tap' as 'a valve with a
free outlet used as a draw-off or delivery point'. By following
the meaning of 'tap' to its logical conclusion, it becomes
apparent that a tap is really a cock, and the modern use of
'tap' is all to cock; a screwdown tap, of the disc component
type, cannot be a tap, or a cock—it should be a valve. Had
the BSI adopted the Scottish cran (qv) or even crane to describe domestic taps the question of the type of valve would
have been irrelevant; a cran may be a cock, tap or valve.
(See also TAP)

VIRREL

A Scottish word, with variations 'virl' and 'virle', for a
ferrule. Although Scottish plumbers pronounce the word
'virrel' they spell it 'ferrule', and 'educated' plumbers tend
to pronounce it 'ferrule' in the English style. But Scots who
look to their OED will find that their 'virrel' is nearer the
mark than they may think; ferrule or ferrell, 1611, was a
corrupt spelling (as if a diminutive of the Latin *ferrum*) of
the older form verrel, verril, from Old French *virelle, virol,*
medieval Latin *virola*. In general terms it means a metal
ring or cap put round the end of a stick or tube, to strengthen
it or to protect it from splitting or wear. In plumbing terms
the proper meaning is used in one or two senses, and in
others the meaning has been corrupted, just as the spelling
was some centuries ago.

Brass Sleeve

While 'ferrule' is the word used in most parts of England and Wales, the phrase common to London and the south east is 'brass sleeve'—a ferrule or virrel in the true sense. The brass sleeve slips over the end of a lead soil or waste pipe and its purpose is to protect the lead in the making of a caulked joint between lead and cast-iron pipes. The correct method of fitting a brass sleeve is to slip it over the pipe until just enough of the pipe shows at the outgoing end to be tafted over the end; thus no waste or soil matter comes in contact with the brass. Many plumbers save a few inches of lead by inserting the lead into the ferrule for a short distance only, and in doing so make a pipe of the brass, which is a poor acid-resisting alloy.

Tailpiece

Although BS uses this as the preferred term for a ferrule or brass sleeve, few plumbers in Britain have ever heard of a tailpiece. Strangely enough, BS uses the word 'ferrule' as the preferred term for two items which bear little resemblance to ferrules. (See FERRULE)

WAESE, WEEZE

A grommet (qv). Heard in Northumberland as weeze, the spelling not known, this is an example of Scottish dialect drifting over the border. Chambers gives 'weeze' as 'a wisp of straw' and 'waese' as 'a small bundle of hay or straw, larger than a wisp'. Presumably the size of pipe to be weezed will decide the plumber whether he wants a weeze or a waese. Again, in Scottish dialect a waese is a circular pad of straw placed on the head for carrying a tub, basket, etc, a straw collar for oxen, a bulky necktie. The plumber's weeze or waese is like a collar of hemp.

WASTE HEAD

In Lincolnshire, a rainwater head when used as a receiver of wastes from sinks, baths, etc. Although Hellyer deplored the use of waste heads, the principle is still widely practised. Many local authorities throughout England and Wales, permit the discharge of waste pipes into rainwater heads inserted in stacks of light cast iron rainwater pipe at suitable floor levels. It is not unusual to see a rainwater stack with a series of rainwater heads at as many as three different floor levels of a building, carrying wastes from two flats on each floor. (See also SOIL PIPE HEAD)

WATER CLOSET

The earliest date given for this term by the OED is 1755: 'A small room fitted up to serve as a privy, and furnished with water-supply to flush the pan and discharge its contents into a waste-pipe below.' Soil pipe is meant. BS 4118, with 'water closet' as a secondary term under 'WC', gives: 'A compartment in which a WC suite is installed.'

However, for very many years the water closet has been understood to be the contraption, or suite, and not the small room, otherwise the British public might not put themselves out so much to find a 'nicer' word than 'closet'; the nicer words or phrases are too silly to enumerate. In the United States of America the water closet is what we know as the 'closet pan'.

Flush Toilet

This term, common with the layman, has also been heard from several plumbers—obviously being polite, but as the word 'toilet' means 'to dress', there seems little connection between 'a toilet' or dressing room and an apparatus called a water closet. Modern closet seats are referred to as toilet seats by makers and public alike, and in this case the term is probably permissible because this type of closet seat is designed—and made snug—for sitting on whilst dressing after a bath, say. However, lift this toilet seat and another seat, with hole, is exposed and this must surely be a closet seat for sitting on while using the water closet—not a toilet seat. In Co Durham an elderly plumber said 'toilet pot'.

L

Office

In recent years 'offices' has been seen on specifications from London authorities. In Scotland 'ofi' has long been a 'common' word for the closet, wet or dry, but usually for an outside 'small room' or privy, and not an indoor bathroom containing a water closet.

Privy

Declaring this word 'obsolescent' BS 4118 describes it as: 'A closet containing a fixed or removable container for the reception of human excrement.' 'Privy' is such a pleasant word that it is a great pity it should be declared obsolete. The OED says of privy: 'A private place of ease, a latrine; late ME . . . that which is secret . . .' What better name for a compartment containing a water closet.

Latrine

BS 4118 contains no information on latrines, so the term is probably obsolete as far as British Standards are concerned. The OED gives latrine as a privy. One usually associates latrines with perhaps a hole in the ground with a perch over it or, at best, a cold damp place with water closets or urinals, and perhaps a place to wash-up, but as latrine is from the same root as lavatory (see US LAVATORY) there seems no reason to associate a latrine with water closets or urinal stalls . . . or privies.

Cloakroom

This, as well as being a compartment in which to keep coats, cloaks and things, is a word used as a cover up by those who cannot bear to say 'closet' or 'water closet'; very often these so-called cloakrooms do not have a hanging hook for a jacket. But there is no better word in the English language for the compartment which contains a water closet, and that becomes apparent when one has a hard look at the 'cloak' part of the word. In Scottish vernacular, 'clock' is pronounced 'cloke' or 'cloak'; there are grandfather clokes and alarum clokes. 'Clock' also means to hatch or sit on eggs, or it is the condition of the hen when she wishes to 'sit'; a clocker is a sitting hen and in Scotland she is a cloker or perhaps cloaker. If one occupies a water closet for too long in Scotland, one is likely to hear a call from without : 'Are you cloaking?' A cloakroom or even a clockroom could well be a room in which one clocks.

Cumming's 'Watercloset'

In 1775 Alexander Cumming, a watchmaker in Bond Street, London, took out the first patent for a water closet, according to the records at the Patent Office; the first patent of any kind in this country was granted by James 1 in 1617. But before 1775 water closets must have been in existence because Cumming describes his patent as a 'Watercloset upon a New Construction' indicating that water closets were well known. It is noticeable that he referred to his invention as the 'watercloset' and his specification included a drawing showing the framework on which the casing (forming the cabinet) was to be built. Illustrations of old types of water closets are always shown in open plan and so it is easy to overlook the fact that the finished article was encased in wood with a hinged cover over the hole; old water closets with mahogany cabinets are still in existence. Further proof of some early type of water closet may be found in the OED

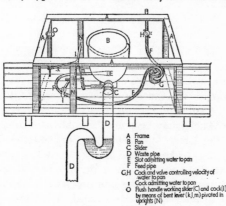

A Frame
B Pan
C Slider
D Waste pipe
E Slot admitting water to pan
F Feed pipe
G,H Cock and valve controlling velocity of water to pan
I Cock admitting water to pan
O Flush handle working slider (C) and cock (I) by means of bent lever (k,l,m) pivoted in uprights (N)

Fig 39 Cummings' water closet (Patent No 1105), 1775

under 'closet': '. . . short for Water-closet 1662' and the OED also states that in 1601 the word 'closet' meant 'cabinet'. (Modern French uses 'le cabinet' to mean a small room, a water closet, etc.)

The receiver shown on Cumming's specification is described as the 'basin' or 'pan' and those two names are still used today; plumbers in some parts speak of closet basins and in other parts of closet pans.

Bramah's Closet

In 1778 Joseph Bramah, of Cross Court, Carnaby Market, Middlesex, a cabinet maker, took out a patent for another

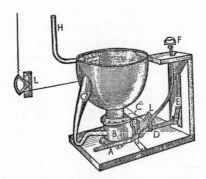

Fig 40 Bramah's water closet, 1778

'Water-closet upon a New Construction' which was a great improvement on Cumming's closet in that some attempt was made to spread the incoming water from the top of the basin, and the valve on the outlet of the basin was a superior contrivance. There may be a few Bramahs still in use, but at the end of the last century Hellyer reckoned that over the years so many makers had modified and improved the Bramah that there was hardly a bit of Bramah left in the valve closets made during the late nineteenth century. Hellyer claimed to have had quite a hand in its improvements, which others were not slow to copy.

Pan Closet

This closet was so called because it had a hinged copper pan which swung down and tipped its contents into the body of the closet, thence into the soil pipe. As the handle was pulled, a wire attached to the lever lifted a plug in the storage tank and the flush of water thus coincided with the tilting of the pan. No one seems to have patented the pan closet; it simply evolved as a cheap answer to the Bramah. As far as is known, the pan closet was first made round about 1790 and was still being made, in spite of condemnation from all quarters, in the early 1890s. There is an old pan closet with cast iron body and copper pan in the writers' small private museum in Ely; it was taken from a local doctor's outhouse in recent years.

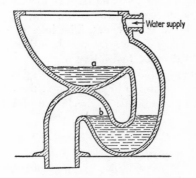

Fig 41   Jennings' "Monkey closet", 1852

Washout Closet

So-called because the flush of water washes out the bowl as opposed to washing down. Josiah George Jennings was the pioneer of the washout closet and he patented what was known as the 'Monkey Closet' in 1852. Like the Bramah and the pan closet, this type has had a good long run; they are still made by Twyfords, a few for use in Britain and others for export. In Co Durham the washout is known as a wash-out pot.

Washdown Closet

The invention of the washdown closet, as we know it today, in 1889, is said to have been claimed for D. T. Bostel of Charing Cross, Middlesex, but who can say that any one person *invented* it. There is little doubt that the washdown, like others, evolved. Sometimes the evolution happened by very slight improvement on a previous design, but by 1880 it had taken its form and awaited only a few refinements like the discarding of the wooden casings and the mastering of the previously wayward flush by changes in the designs of flushing rims.

It has been difficult to get any information on the origin of the long hopper and its mate the short hopper of the nineteenth century—Hellyer referred to them as being old-fashioned—but before the long hopper there was one of the simplest forms of washdown in the last century, which Buchan calls the cottage closet. The inlet of the lead trap was opened out to receive a hopper-shaped basin and it was supplied by a 'valve-cock' with lever.

The 1870s saw a hastening of the evolution of the washdown closet, thanks to the highlighting of the subject of sanitation caused by the illness of the Prince of Wales in 1871.

Carmichael Washdown Closet

The Carmichael washdown closet patented by Buchan in 1879 gave the shape of things to come and, with hindsight, one can see it as evolving from an idea which a bright plumber had back in 1853 (see BUCHAN TRAP). Whether Buchan knew it or not who can say, but when he invented his two-piece Buchan Trap in 1875 he had the basis for his Carmichael closet; the bottom half of the trap—ie the water section—merely had to be modified with a base and a greater area of exposed water surface to give him the bottom half of his closet, which was later to be made in one piece or in two pieces. Although the Carmichael shows a rim, it served as a kind of anti-splash only, the flush coming from the then orthodox metal fan. Davies said that this closet was '. . . what may be called the acme of hopper closets'.

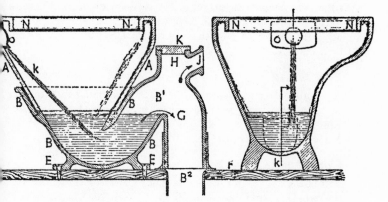

Fig 42 Buchan's Carmichael closet, 1879

WATER HAMMER

A knocking noise in water pipes caused by high water pressure usually combined with a fault in the plumbing system.
It may start with a low, rhythmic knocking which increases
to a hammering, quite violent at times. The turning off of a
tap may be sufficient to cause an initial shock in the pipes
and then the knocking begins.

Housewives may insist that there is air in the pipes and
the plumber who tries to explain that air cannot get into
such pipes is wasting his time (see AIR LOCK).

The causes of water hammer are sometimes difficult to
locate and once found are not always easy to cure. If it
occurs in a new system, a search must be made for loose
pipes; two pipes in contact with one another can cause the
hammer. Copper tubes should have an abundance of fixing
clips to hold them tight. The shell-like construction of gas
water-heaters induces the hammering when the water supply
is off the main, and the gas flame may flare alarmingly until
the hammer stops. Air vessels should be fitted on new
work.

If the fault develops on existing work, a faulty washer on
tap or ballvalve is most likely to be the cause; amateur

plumbers frequently fit very soft washers on taps, whereas hard washers are preferable. Another cause is a loose spindle on a tap; repacking of the gland (see US JOURNAL) can be a remedy.

Chattering

A suitably descriptive name for water hammer heard in the London area and several places in the southern half of England. Certain sounds are not so much a 'hammer' as quick tapping or chattering.

Concussion, Reverberation

Secondary names for water hammer given by BS 4118, but the working plumber is unlikely to use 'reverberation' or even 'concussion' when speaking of water hammer.

WATER HEAD

A rainwater head. Heard in Llandiloes, Wales.

WATER WASTE PREVENTER

Labelling the term 'obsolescent', BS 4118 describes it as meaning: 'A flushing cistern of the siphonic type.' For some years, 'water waste preventer' has been a term peculiar to the London area and parts of the south of England, and is still used today. When referring to closet cisterns plumbers in those areas speak of 'WWPs'.

Although the phrase could well be applied to any appliances designed to save water, it has been reserved for closet cisterns. But the water waste preventer has not always been of the siphonic type of flushing cistern.

Early water closets were supplied from a storage tank, which also supplied water for domestic use. The pull of a wire or cord on a lever raised a plug—sometimes called a drop-valve—in the tank and if the pull was held long enough, the tank could be emptied: there was nothing of the waste preventer about it.

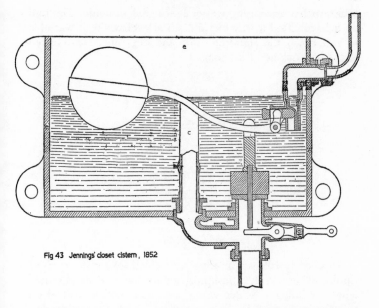

Fig 43 Jennings' closet cistern, 1852

Some time in the nineteenth century a double cistern evolved. An elderly plumber in Bristol remembered it well and recalled the term 'double decker' for the double cistern. It was really an ordinary storage cistern with a partition to form a smaller compartment. There were many types of valves used in attempts to find an effective way of saving water, but the double cistern was dropped during the last quarter of the nineteenth century when the closet cistern was beginning to take shape.

In 1852 Jennings patented a small cistern—he did not call it a water waste preventer—which gave the shape of many closet cisterns for years to come. When the chain or wire was pulled, a forked rod rose in the cistern and held up the ballcock (qv) whilst the flush was executed, and so no water could enter until the pull was released and the outlet of the cistern was closed. The use of the siphon had not been thought of at that time.

John Shanks of Barrhead, Renfrewshire, patented in 1871

a cistern for preventing waste. It was not siphonic. The pull on a lever lifted a tray of water to a level where it was able to run down the standpipe into the flushpipe. He referred to the ballcock as a float cock.

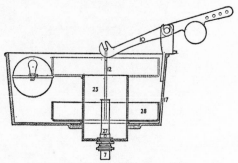

Fig 44 John Shanks' closet cistern, 1871

Apparently Hellyer in 1875 was still thinking in terms of tipping water down a flushpipe, for in that year he patented what he called a 'water waste preventer' which worked on the tipping principle. The specification of his patent describes the action :

. . . consists essentially of a tipping or tilting cistern hung upon centres within an outer fixed receiver or container, one of the said centres consisting of the hollow plug of a cock, which is fixed to the inner side of the outer cistern and works water-tight within a socket forming the body of the cock, which is brazed or otherwise secured to the inner side of the inner cistern . . . The act of tipping or tilting the inner cistern on its centres of suspension for

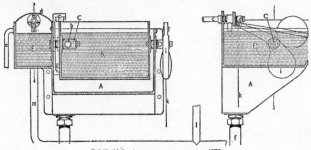

Fig 45 Hellyer's water waste preventer, 1875

the purpose of discharging its contents into the outer fixed receiver or container effects the immediate shutting off of the supply to the said inner cistern, and hence no waste of water ، ؛ ؛

Thomas Crapper of Chelsea and many others specialised in *perfecting* the siphonic cistern and by 1887 D. T. Bostel and W. Bostel patented 'Improvements in Syphon Flushing Cisterns'.

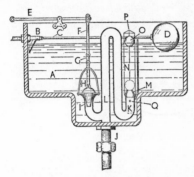

Fig 46 DT & W Bostel's syphon flushing system

WEDGE BIT

A hatchet bolt (qv). An apt, descriptive name, heard in Co Durham, for this soldering tool.

WEEP

When a joint in a water pipe has a very slight leak, barely enough to drip, it is said to weep, or sweat. As a noun, the drip is a bead, a sweat, a tear, or a weep; any of those descriptive words may be heard throughout the British Isles. In the case of solder joints, plumbers do not worry unduly about a very slight weep as the lime or sediment in the water supply is almost bound to seal the pore in the solder. A tap with a small hammer soon stops a weep.

Sweat

A water pipe dripping with condensation is said to be sweating, and the plumber may find it difficult to convince a client that this kind of sweat is not, in fact, a weep from a faulty joint. A case comes to mind when a client insisted that there must be a leaking water pipe in a plastered wall; there were no water pipes in that part of the house. Removal of some cheap wallpaper revealed gloss paint on the plaster; some decorator had cleaned out his paintbrush on the wall, with the result that, in humid weather, a dampness which would normally be absorbed into the plaster showed as a damp patch on the paper. The client insisted on a second opinion.

WEEP-HOLE

Heard in the south of England with reference to 'bleeding' a water pipe. (See also PIN-HOLE)

WELL

Scottish vernacular for a cran (qv) or a tap (qv). The name probably comes from the hole-in-the-ground type of well, from which several householders could draw water; as a plumbing term in the nineteenth century it was the cran set up in a back court or a square, sometimes just at the roadside for the use of householders and passers-by alike. Kennedy's self-closing well was a round iron case with an iron knob, for a handle, projecting from its face. The knob handle, which was attached to a weighted spindle, was turned to open the cock and closed, by virtue of the weight, when the handle was released.

WELT

To welt is to join together two sheets of metal by folding the edges, engaging the folds, and either pressing together in a vertical position or dressing down flat. An upright welt is called a standing seam. Either single or double folds may be made, the completed joint being termed either a single welt or double welt.

WIRE BALL

A Scottish term with variations 'wire ball grating' or, simply, 'ball grating', for a ball-shaped wire guard inserted into the top of a ventilating pipe to prevent birds from nesting in the pipe, or into a gutter outlet as a safeguard against stoppage due to silt and leaves. In England and Wales, it may be called a wire balloon or balloon guard (qv). They may be of ordinary galvanised wire or, more expensive, of copper wire.

YARNER

Gaskin (qv) is known as yarn or spun yarn in the greater part of England and in Wales; thus a yarner or yarning tool is used to pack the yarn into spigot and socket joints. (See also STEMMER)

Supplement of terms used in the United States of America

Note: Cross references (See . . . and See also . . .) are to terms included in this Supplement unless preceded with 'UK', when they will be found in the main body of the text.

ALL-THREAD

A running nipple. (See UK BARREL NIPPLE)

BACK FLOW PREVENTER

Self-explanatory term for a reflux, non-return or back pressure valve.

BALLCOCK

This misnomer is perpetuated in the United States just as it is in Britain, 'ballvalve' being used as an alternative.

BATH TUB

Or simply 'tub', for what is known in Britain as a bath. The term conjures up almost forgotten memories of Saturday night baths in a wooden tub in front of the kitchen fire; the tub was known then, in Scottish dialect, as the bine, boin, or boyne. There are indications that 'bath tub' is creeping back to Britain.

BELL

As in bell-and-spigot, meaning spigot-and-socket, or, as the Scots would have it, spigot-and-faucet. 'Bell' is seldom used by itself (see also HUB) in relation to cast-iron pipe joints.

BEND

A formed bend. That is, a pipe bend made by the plumber or pipe fitter on a bending table as opposed to a manufactured fitting. Occasionally, this formed bend is referred to as a long radius 'ell', but, to be correct, it is simply a bend.

BIBB

May also be spelt with one 'b' as in the British bib tap (qv). The word is frequently used among plumbers meaning a faucet (American style, not Scottish). Unlike the British bib, the US term does not only mean a tap at a sink; a faucet fitted outside for a hose connection becomes a 'hose bibb'.

BLOWOFF

This is not a safety valve as the British plumber might think from the name. It is an outlet, controlled by a valve, on a boiler or hot water storage tank to permit emptying or discharge of sediment. The British plumber would call it a draining tap or a sludge cock.

BUFFER PIPE

An air vessel. It will be noted that the US 'buffer', like the Scottish 'air cushion', gives the proper sense of the air vessel's purpose.

CALK

Or caulk. The meaning is the same as the British caulk (qv). The old English spelling 'calk' is perpetuated in the United States: 'calking tool' is also 'caulking tool'.

CENTER SET

The faucet for a bath tub is a 'bath center set' and the one for a lavatory is a 'lavatory center set'. In Britain, 'combination taps' is common usage but BS 4118 gives 'combination tap assembly' with 'combined fitting' as a non-preferred term.

COCK BOX

A stopcock box; also a reminder of the unconventional Scottish toby. (See UK TOBY)

CROTON

In New York City the branch hot and cold water pipes from the risers to the fixtures take their name from New York's earliest reservoir, the Croton Reservoir. The branch pipes are Crotons.

CROW'S FOOT

Whereas this is a basin spanner in Britain, in American terms it has been described as the plastic washer used in conjunction with a backnut in fixing a faucet or center set. It has been said that the crow's foot is so called because of the spidery veins across the surface—like the marks made by a crow's foot.

DEAD MAN

One unverified source called a turnpin (see UK TURNPIN) a dead man. No reason could be given for the name.

DECK FAUCET

The faucet at a kitchen sink. Enquiries in the United States

M

about the 'deck' part of the term have not met with success. However, Pegler Ltd, Brassfounders of Bury St Edmunds, Suffolk, England, and makers of 'Prestex' compression-type fittings for copper tube, show in their illustrated price list a 'deck sink fitting' and also a 'deck pattern basin fitting'; the same Prestex list shows pillar sink fittings as well.

The pillar fittings have hot and cold supplies which come through the sink top with two separate pipes with flanges or escutcheons. The bridging pipe with gooseneck outlet is above the surface of the sink top. In the case of the deck pattern, no individual supplies are shown, the controls and gooseneck being incorporated in what might be called an elongated escutcheon—the 'deck'; the deck is about one inch in depth.

Thus, while the term 'deck faucet' might mean 'sink faucet' in the United States, decked fittings could also be used for basins and bath tubs.

DOPE

Jointing paste for screwed joints on pipes.

ELL

It would seem that although this is said to be short for elbow (see UK ELBOW), ell, through common usage, has become a word in its own right. It also means what is known in Britain as a bend (manufactured); there are short radius ells and long radius ells.

FAUCET

The OED quotes Knight: 'The enlarged end of a pipe to receive the spigot end of the next section'; that would be middle nineteenth century. The term is not used nowadays for the enlarged end of a pipe, but it has been revived with the meaning 'tap' and it is interesting that the very early faucet was a tap for drawing liquor when drinking or 'bibbing'. (See also UK BIBTAP, BIBB and UK FAUCET)

FLUSH TANK

Or toilet tank. A closet cistern in UK (qv).

GOOSENECK

A faucet with a long swinging arm.

HICKIE

The bent bolt was called a hickie by one informant who described it as an electrician's tool. So far there has been no verification of the term.

HOSE COCK

In the United States, as in Britain, there is some disregard as to whether a valve is a cock or not. Hose cocks are usually screwdown valves and not cocks. (See also BIBB)

HUB

A name for what is known in the UK as a socket, or a faucet, of a cast iron pipe. The 'hub and spigot' is also known as the 'bell and spigot'. A correspondent from the Cast Iron Soil Pipe Institute says of the term 'socket': 'The copper people talk about a pipe socket in a fitting (ie a fitting for copper tube). We call it a shoulder on which the pipe fits snugly down in the hub, formerly called bell.'

HUSH PIPE

A gentler and more descriptive name for the harsher British 'silencer tube' in a closet cistern. It screws into the ballvalve and directs incoming water below the surface of water already in the cistern. It does not silence, merely reducing the noise of rushing water to a gentle 'shush'.

INCREASER

A reducer in UK. A pipe fitting for connecting together two or more pipes of different diameters.

JOURNAL

More of an engineering term, it is of interest in that 'journaling' has been used in a description of faucets (qv) in an American plumbing journal; in machinery, the journal is the part of a shaft or axle which rests on the bearings.

Journal Box

The box which contains the journal and its bearings. Plumbers in Britain would call it the stuffing box or packing gland when dealing with taps, valves, etc.

Journaling

Fibrous material with which the journal is packed—saturated with oil or grease. When packing a stuffing box, or packing a journal, the plumber will resort to a piece of soft string or cord which he saturates with tallow. Having been packed, the material is held tightly in place with a packing gland ring which screws into the top of the journal box.

LAVATORY

A gentleman, a layman, whilst discussing plumbing with the author, remarked: 'It makes my flesh creep to hear people call the lavatory a toilet.'

The author replied: 'It makes *my* flesh creep to hear a water closet called a lavatory.'

American and British laymen may mean 'water closet' when they say 'lavatory', but in American plumbing circles 'lavatory' or 'lav' means wash basin. Bath boutiques in the United States advertise lavatories or lavs, and even vanity lavs when referring to wash basins. A few plumbers in Britain refer to the wash basin as the lavatory basin but BS 4118 gives 'lavatory basin' as a non-preferred term for a wash basin.

The word lavatory is from the Latin *lavatorium,* a place for washing, and, incidentally, latrine is also from the Latin *lavare,* to wash.

LEADER

Or conductor, or down spout; a rainwater pipe. Plumbers in some parts of Britain will recognise 'conductor' and/or 'down spout' as having the same meaning in their regions. But although one source of information from the United States is quite definite that 'leader' is synonymous with 'conductor', the US National Plumbing code is puzzling in that it refers to 'inside conductors' and 'outside leaders' and leaves one uncertain as to whether a leader becomes a conductor when it is inside a building, or if leaders and conductors serve two separate functions. In the early part of the twentieth century Starbuck juggled with 'leaders' and 'conductors' in the most confusing manner, but even he does not make clear to a mere reader of American plumbing the difference between the two.

However, another informant says that in some parts of the United States a leader is an outside rainwater pipe and a conductor an inside one, and in other parts the leader is inside and the conductor outside.

Clarke, a London plumber and employee of S. S. Hellyer near the end of the nineteenth century, writes about rainwater leaders, but Clarke wrote for the American market and would therefore use American terms.

The OED tells us that a leader is someone who leads, or something which leads, and does not make any specific note about pipes leading. But Chambers gives: 'a pipe for conveying water'; 'leader' is not a Scottish plumbing term, although given in the general sense.

LINER

This word has been given in the phrase 'storm water vertical liners' with reference to leaders or conductors, whether they are inside or outside the building.

LOCAL VENTING

A method of ventilating by means of a vent pipe taken from the fixture side of a trap to the open air; the inside of the fixture as well as the room is thus vented. Closet pans for such a system, which is becoming obsolete, used to have a side horn or nozzle—the Scots would call it a pap (see UK PAP)—connecting the vent pipe (see also SPUD). A similar system, used in Britain, although it was not known as 'local venting', is described by Buchan who advocated that the top plate of the cast-iron case of a pan closet (see UK WATER CLOSET) should have two holes drilled in it for the connection of pipes to ventilate the space between the copper pan and the trap; fresh air was supposed to enter through one pipe, and the foul air exit through the other!

MAUL

It has been pointed out that plumbers in the United States do not fit flashings or gutters on roofs; roofwork is done by the sheet metal worker. However, the mallet which the British plumber uses for lead work is known as a maul in America. The word goes back to the English medieval period and is from the Latin *malleum,* a hammer.

OAKUM

Gaskin in Britain (qv). Many British people associate oakum with the United States but it has its root in Old English, meaning literally 'of combings'. The picking of oakum, ie teasing out old rope, for use in caulking of ships' seams, was the employment of convicts and workhouse inmates as far back as 1481, according to the OED.

PEAN

The pane of a hammer. One is tempted to think that this is a quaint American way of spelling 'pane' but perhaps there is a close connection between this American 'pean' and the Scottish 'peen' or 'pean'. (See UK PIEND)

P-TRAP

Although P-trap has replaced the old form $\frac{1}{2}$S-trap in Britain, both terms are used in the United States, as is $\frac{3}{4}$S for a Q-trap (see UK TRAP)

ROOF-JACK

Or roof fitting, or roof flashing, or vent flashing. A lead slate in Britain (qv). A jack, says the OED, is a short and close fitting jacket; the roof jack is quite short and fits closely round the vent pipe as it comes through a roof.

SILL COCK

A hose bibb (see Bibb). One can only suppose that this outside hose bibb is usually in the proximity of a windowsill or even fixed to a sill.

SPAGHETTI

An unverified source gives this as annealed (soft) copper tube of small bore. So called, perhaps, because of the spaghetti appearance of a coil of annealed tube and the ease with which it can be bent by hand.

SPUD

The integral nozzle or horn on the side of a closet basin designed for local venting? Starbuck shows a spud as a nozzle. Informants in the United States have no knowledge of such a plumbing term, but the story of the spud has been found nearer home, in England; Twyfords of Stoke-on-Trent, Staffordshire, connect the spud with a local potter of the nineteenth century.

Thomas Maddock, of a family who had been potters for several generations, left England in 1847 and went to America. He was the first successful manufacturer of sanitary ware in the United States. About 1880 he was granted a patent for his 'spud' inlet device which, Twyfords say, differs very little from the one used today, both in the United States and in Canada, as the standard *inlet* connection to a closet pan. As far as can be ascertained from the illustration supplied by Twyfords, the 'spud connection' is a brass fitting comprising a socket on one end for receiving the nozzle and on the other end a male thread for connecting a pipe. While the spud connection may be an 'inlet device' today, it was certainly an outlet device when used for local venting, the current of air being from the closet. Twyfords' illustration calls the nozzle a closet 'basin vent' and that raises the question of whether the nozzle or the brass connection is the spud. In the general sense, a spud is someone or something short or stumpy; the spud connection is short or stumpy, but the nozzle on the pan is also short and sticks out from the pan like a stump. The term 'spud connection' suggests that the connection is for connecting to a spud, so it is likely that Starbuck is correct in showing it as part of the closet.

Twyfords used to supply closets with vent spud to the Canadian and certain European markets, but there is no longer a demand for them.

STREIT ELBOW

A street elbow in the UK (qv). A retired American plumber says that in his early days at the trade, the street elbow was known as the streit and it was thought to be the name of the inventor of this fitting. This term could probably do with further research.

TELL-TALE PIPE

A warning pipe, ie an overflow pipe of a tank, flush tank or storage tank, so fitted that its outlet is in a conspicuous position where any discharge will be noticed.

TRAY

A word which rather suggests a shallow vessel, but in the

United States the deep section of a combination tub and sink, known as a tub in Britain, is called a tray. For example, tub and sink becomes sink and tray. A double sink is also a sink and tray, the tray part may have a garbage grinder fitted to the waste outlet.

WALL HYDRANT

A hose bibb (see BIBB). BS 4118 gives, for hydrant: 'A device, by means of which water may be discharged from a pipe for extinguishing fire . . .' The word 'hydrant' simply means an apparatus for drawing water directly from a water main; that is the meaning of the American wall hydrant. A garden hose bibb can be a wall hydrant.

List of Sources

1. BOOKS

Buchan, William Paton. *Plumbing* 6th edn 1892, 7th edn 1897, Crosby Lockwood, London

Clarke, J. Wright. *Plumbing Practice* 2nd edn, the Engineering and Building Record, New York (London, 1891)

Davies, Philip John. *Standard Practical Plumbing,* vol 1, 5th edn 1905; vol 2, 2nd edn 1905; vol 3, 1904. Spon, London, and Spon & Chamberlain, New York

Hellyer, S. Stevens. *Lectures on the Science and Art of Sanitary Plumbing,* 2nd edn 1883, Batsford, London

Hellyer, S. Stevens. *The Plumber and Sanitary Houses,* 3rd edn 1884, Batsford, London

Herring-Shaw, A. *Domestic Sanitation and Plumbing,* 1st part 1909; 2nd part 1911, Gurney and Jackson, London

Mitchell, Charles F. *Building Construction,* Part 2 1919, Batsford, London

Reyburn, Wallace. *Flushed with Pride,* 1969, Macdonald, London

Starbuck, R. M. *Modern Plumbing Illustrated,* 4th edn 1922, the Norman W. Henley Publishing Company, New York

Weaver, Lawrence. *English Leadwork, Its Art and History,* 1909, Batsford, London

Wright, Lawrence. *Clean and Decent,* 1966, Routledge & Kegan Paul, London

Woolgar, W. J. (ed). *The Practical Plumber and Sanitary Engineer,* 1946, Odham's Press, London

Manas, V. T. (ed). *National Plumbing Code Handbook,* 1957, McGraw-Hill, New York

The Plumber's Handbook, 1968, Lead Development Association, London

2. DICTIONARIES AND ENCYCLOPEDIAS

Concise Oxford French Dictionary, 1966, Clarendon Press, Oxford

Dictionary of Slang & Unconventional English, Partridge, Eric, 1937, George Routledge & Son, London

Dictionary of the Vulgar Tongue, Grose, 1785 (cited by Partridge)

Encyclopaedia Britannica, 24 vols, 1929, 14th edn

The Penguin Dictionary of Architecture, 1969, Penguin Books, Harmondsworth

Scots Dictionary, W. & R. Chambers, 1965, Edinburgh and London

Shorter Oxford English Dictionary (2 vols) 1936. 2nd edn

3. CATALOGUES

Allied Ironfounders Ltd. *Cast Iron Goods,* 7th issue

Britannia Iron & Steel Works Ltd. *Handbook*

Carron Company. *Cast iron goods,* 1955

Crane Ltd (England). *Malleable iron pipe fittings,* 1967

Dawson & Co Ltd. *Drainage goods,* 1962

Frazer & Ellis Ltd. Catalogue no 65, *Sanitation*

Greenwood and Hughes Ltd. *Grevak Monitor leaflet,* CM4e, 1969

Grey & Marten Ltd. *Catalogue,* 1923

Howie, J. & R. Ltd. *Catalogue,* SfB Ig UDC 696, 133

Kay & Co (Engineers) Ltd. *Catalogue of Pipe Fittings,* 31st edn

LeBas Tube Company Ltd. *Malleable Iron tube fittings,* publication no LB/12

McAlpine & Co Ltd. *Catalogue,* SFB 52 UDC 696, 129

Max Sievert A. B. Catalogue no. 270, *Blowlamps for petrol*

Metal Agencies Co Ltd. *Catalogue* no. 46, 1927, Bristol

Prestex Compression Joints. *Catalogue* 1969 edn
Shetack Tool Works Ltd. *Catalogue of Plumbers' Tools*

4. MISCELLANEOUS

Armitage Ware Ltd. *Instruction leaflet F1/WC*
Associated Master Plumbers & Domestic Engineers. *List of Tools required to be provided by plumbers,* 1966
Building Materials Market Research Ltd. *National Price List,* 1964, 1966
Cast Iron Soil Pipe Institute. *Cast Iron Soil Pipe & Fittings Handbook,* 1970, Washington, DC (Library of Congress Cat Card no 67–21122)
GKN Screws & Fasteners Ltd. *GKN Handybook for Householders,* 1970
The Journal of Plumbing, Piping & Hydronics, 1970, New York
Marley Tile Co Ltd. *Small diameter discharge pipes in dwellings,* G. J. W. Marsh, 1969
Bohn's Illustrated Library. *Pictorial Handbook of London,* 1854
Proceedings of the Philosophical Society of Glasgow, vol 27, 1895–96, pp 151–5. 'The late Mr W. P. Buchan', James Chalmers

5. PATENT OFFICE SPECIFICATIONS

D T & W Bostel
    14955 of 1877   Syphon Flushing Cistern
Joseph Bramah
    1177 of 1778   Water Closet
    3611 of 1812   Constructing laying of pipes, etc
W. P. Buchan
    1499 of 1875   Ventilating water-traps
    1662 of 1875   Water Closets
    2102 of 1877   Ventilating appliances
    2745 of 1878   Ventilators
    1853 of 1879   Instruments for measuring and indicating thicknesses or weight of sheet metal plates, etc

    4717 of 1879   Ventilating buildings, carriages, etc

    5272 of 1879   Water closets : drain and other pipes

Alexander Cumming

    1105 of 1775   Watercloset

S. Stevens Hellyer

    2620 of 1875   Preventing waste water

    4424 of 1876   Receptacles for sewage

    4913 of 1878   Water closets

J. G. Jennings

  11728 of 1847   Water closets making joints and con-
                       nection of pipes

  12012 of 1848   Cocks or taps for drawing off liquids
                       and gases

  14273 of 1852   Water closets, traps and valve pumps

A. F. Shanks

    3188 of 1875   Water closets

J. Shanks

    1492 of 1870   Water closets

    2968 of 1871   ,,   ,,

    2643 of 1872   ,,   ,,

     265 of 1879   ,,   ,,

6. BRITISH STANDARDS INSTITUTION

Glossary of Terms applicable to Roof Coverings, BS 2717,
1956

Glossary of Sanitation Terms, BS 4118, 1967

Acknowledgements

So poor is my memory and so many are the people who have offered help and hospitality to me during the last four years that my list of acknowledgements can name but a comparative few of my helpers. Miss Una Long has worked with me from start right to time of posting the proofs back to the publisher.

My thanks to R. V. Cooper, MBE, CGIA, FIOP, MIPHE, when he was secretary of the Institute of Plumbing, and to H. Ryland, FIOP, editor of the *Plumbing and Heating* journal, for helpful discussion and suggestions.

A few of the individuals who offered more co-operation than one might ordinarily expect from commercial organisations were Scott Arbuckle, United States Brass Corporation, Texas; F. Hance, Crane Ltd, London; R. H. Emerick, consulting mechanical engineer, North Charleston, USA; T. McCarthy, plumbing technical assistant, *Contractor* magazine, Massachusetts, USA; A. E. Scherm jnr, Crane Co, New York; Seth Shephard, editor-in-chief, BP *Contractor* magazine; J. Young, Howie-Southhook Ltd, Kilmarnock, Scotland.

Then there was Miss Christine Fanning, Branch Librarian, Ely Public Library, who obtained for me books I had never hoped to see, and also John Winder of *The Times* and David Woodhead of the *Sunday Telegraph* whose articles on my research brought me numerous contacts.

Also helpful in supplying documents, drawings, etc, were the Cast Iron Soil Pipe Institute, Washington DC, USA; GKN Screws & Fasteners Ltd, Warley, Worcestershire;

Greenwood and Hughes Ltd, Littlehampton, Sussex; Mitchell Library, North Street, Glasgow; National Library of Scotland, Department of Printed Books, Edinburgh; Price's Patent Candle Company Ltd, London; Primus-Sievert AB, Sundyberg, Sweden; Twyfords, Stoke-on-Trent.

My thanks to all plumbers who took time to converse with me in the course of my fieldwork.

Index

compiled by Una Long

N*